# Frameworks, Artworks, Place

# Consciousness Literature & the Arts 11

# Frameworks, Artworks, Place

## The Space of Perception in the Modern World

EDITED BY

TIM MEHIGAN

Rodopi

Amsterdam - New York, NY 2008

Cover image: James McArdle: Gush (2007). Chromogenic print.

Cover Design: Aart Jan Bergshoeff

The paper on which this book is printed meets the requirements of "ISO 9706:1994, Information and documentation - Paper for documents - Requirements for permanence".

ISBN: 978-90-420-2362-8
ISSN: 1573-2193

Printed in the Netherlands

# TABLE OF CONTENTS

# The Space of Perception

Tim Mehigan

The question of space has received attention in recent approaches to consciousness. This has partly arisen as alternatives have been sought to traditional cognitive science, whose material focus on the mind's processes appears to account for only a small part of what we understand consciousness to be. It is also due to the rise of phenomenological and psychological approaches to the mind, which ascribe importance to the data flowing from the subjective standpoint and thus make the way consciousness is shaped by what might be called the 'aura' of perception[1] a key moment of investigation. These approaches have their source in the 'Copernican turn' towards the subject inaugurated by Kant's philosophy in the late eighteenth century, although little significance was initially found in this emerging idealistic focus on the mind beyond certain stirrings in German art and philosophy, and its potential utility remained largely unnoticed outside Germany for more than a century. It was not until the neo-Kantian revival in the late nineteenth and early twentieth century that the subjective dimension of consciousness commanded attention once more and to some extent entered the European mainstream. Edmund Husserl, one of the most prominent thinkers in this revived tradition, made the link with Kantian notions of the mind and the break with empirical psychology quite explicit in his new phenomenology (Husserl I 1913: 214).

The question of space in approaches to consciousness has progressively moved to the fore since the pronouncements of early phenomenology. Heidegger, a student of Husserl, attached importance to the perspectivalism of perception, where the world is not just observed in pictures made by the mind, but the pictures of the mind, from a

[1] Following Benjamin, who spoke of the 'aura' of the artwork (Benjamin 1983).

certain point of view, in fact make up the world. Key statements in Wittgenstein's early philosophy affirmed this interest in the pictorial aspects of perception and the role of mental space in enlivening them. The interest in perspectivally generated consciousness, which was obliged to assume a dynamic subject in whom acts of consciousness issued, was paralleled during the same period by new developments in science. The insight of experimental science that scientific attention is by no means neutral in its access to the external world, but delivers data specific to the type of examination undertaken,[2] notably at the quantum level of atomic physics, has penetrated ever more deeply into the natural sciences. As the understanding of what constitutes scientific objectivity was updated in an age of quantum mechanics, so the question of subjectivity has bulked ever larger for all parts of the scientific enterprise. The once sharply drawn distinction between the sciences and the arts has also been open to reassessment. It is now possible to estimate more clearly the reciprocal exchanges that flow between the 'two cultures' (C.P. Snow), and to see – as the essays in this volume do – a 'preobjective' patterning in the insights delivered by science. The contributions to this volume, which betray perhaps familiar connections with the Aby Warburg-Erwin Panofsky tradition of art history as well as to phenomenology, for this reason also follow more innovative directions in fields as diverse as cognitive neurobiology (Maturana, Ramachandran) and linguistic philosophy (Lakoff / Johnson), where the question of situated, embodied consciousness has been given new importance.

A significant leap forward in our understanding of the nature of perception occurred somewhere between the sixteenth and the eighteenth centuries. A distinction was introduced – let us say it began with Descartes – between the process of ordering experience through the activity of the mind, on the one hand, and the existence of one's physical body in 'outside' space, on the other. The ordering activity of the mind, where 'units' of perception were created from what could be taken in of the world through consciousness, separated the mind's processes from what could be assumed about the make-up of the world outside the boundaries of the self. That the mind required an idea of space to enable its processes to function in consciousness, and

[2] As noted, for example, by Max Planck in 1933: 'Every measurement first acquires its meaning for physical science through the significance which a theory gives it' (Planck 1933: 92).

that this idea of space is apparently different from the space of the world 'out there', did not properly emerge until Immanuel Kant separated 'transcendental' awareness from the 'thing in itself'. Kant was only in a position to draw this distinction between outside space and the mental space of consciousness because of Hume's work on empirical causality immediately before, which had identified a discrepancy between the impressions human beings received of the physical world and the constitution of that world.

This moment of separation of mind from the space of the outside world was of enormous utility for the emergence of the new sciences. For the first time, human awareness was able to probe the physical environment in which human life was cast, at least notionally free from the encumbrance of superstition and mind-created fancies. Francis Bacon, writing shortly before Descartes, spoke of the advantages that would accrue to human beings if they were able to release themselves from certain 'idols' – certain prejudices affecting understanding – that stood in the way of direct perception of the physical world. Galileo Galilei, training his telescope on the heavens in order to test Copernicus's hypothesis about the heliocentric nature of the then known universe – Copernicus had reached his conclusions on the basis of unaided observation and mathematical calculation alone – offered a vivid example of this new science in action: its insights, derived from the technical means of human invention, appeared self-evident to the human eye, its truth therefore irrefutable. To be sure, the new science had not yet unfolded its methodology in more than rudimentary ways. Descartes sprang into the breach with his *Discourse on Method* of 1637, setting out a series of simple steps by which scientific truth could assert itself. The method he advanced, which was closely aligned with Bacon's treatise on scientific procedure in the *Novum organum* (1620), involved a type of eliminative induction that, as Fred Wilson has shown, confirmed the innate preferences of the medieval mind, that is, it confirmed the drift towards Aristotelian logic and the drift away from Platonic models of understanding, even as it progressed beyond both in a certain sense. Thus was inaugurated the now familiar feel of modern science: it is Aristotelian in its outward focus on physical 'extension' and in its incrementally gathered inductive truths, while human awareness still roams Platonic terrain in its speculations about the spiritual condition of human beings that appears inseparable from the circumstances of mental life. Since modernity involves a compact – a type of 'Burgfrieden' or truce – between the demands of the new science and its ethic of progress on the one hand, and the feeling of spiritual connectedness that human beings sense with other human beings on the other, modern consciousness has been obliged to tolerate the Cartesian gap between mental and physical life, between what the spirit intuits about the human condition, both for the practical purposes of living and purely speculatively, and what science asserts about the nature of the physical environment into which the human body is inserted or 'extended' and on which it impinges.

This gap between mental and physical life – between the 'eternal' ideas of Plato and the practical truths of Aristotle, which was opened still further by Luther's Reformation – might not have survived in its peculiar form but for an entirely unheralded development in human history: the rise of a new kind of art. Scholarship is still to exhaust the significance of what has come to be known as the Renaissance. Nevertheless, it was launched by the solution of a single problem that appears to have eluded the ancients almost completely: the problem of how to depict the position of figures on a flat planar surface so as to render their relation to each other in more or less accurate ways. Until the discovery of the technique of single point perspective around 1430 (first described by Alberti, who drew out transversal lines from an imagined point of observation to objects in space, slicing through this space vertically at a given point to construct the cross-section of a 'visual pyramid'), art was obliged to defer to the achievement of innate human awareness in rendering the third dimension. The Renaissance changed all this. On the basis of a mathematical idea – for mathematics is needed to draw the orthogonal lines and geometric relations that obtain between the objects depicted on the flat surface – art could suddenly produce a representation of objects in space that conformed more or less to the way these objects seem to appear to human perception. Over the entire period from the Renaissance to the Enlightenment, the representation of objects achieved by single point perspective was taken to be the actual way objects are given to consciousness in intuitive awareness. We now recognize limitations to this view. Nevertheless, art's new found capacity to paint realistic pictures of figures and objects in space was one important reason it rose to prominence during this period and could lay claim to a certain kind of transcendence. Human awareness soared on visions of the sensuous body depicted in a space enlivened, in the manner of the great artists of the Renaissance, in ways that aligned with both spiritual and corporeal truth. Art therefore began by stages to bridge the gap that the new Protestantism was deepening between religious understanding and sensuously intuited individual life. In this early modern view, art was nothing that existed for its own sake; art rather claimed attention as an envoy from the realm of transcendence.

The idea that philosophy might have followed this same coming together of the real and the spiritual in the symbolic form of Renaissance art, as Panofsky suggested in a groundbreaking study, is one of the insights developed in this volume of essays. The background to Peter Leech's investigation of how perspective came to be a branch of philosophy, culminating in the ideas of Kant, may be sought in Kant's notion of the appearances. The appearances was a notion Kant deployed in the wake of Hume's analysis of causality in order to resolve the problem of how the body was held in the 'absolute' space postulated by science – the new science of Descartes, Bacon and Galileo – yet at the same time was also connected to the free space of the mind, which was not pinned down to the same corporeal verities. Kant's solution was to suggest that objects *appeared* to the outer senses, which is to say that they *only* thus appeared in ways that showed a separation from human percepts and fell short of suggesting knowledge of their true nature. How objects were really constituted – a problem Kant introduced with references to 'the thing-in-itself' – thus remained beyond human ken in an absolute sense. This, the first insight of Kant's philosophy, was a necessary concession to Hume's scientific scepticism – a scepticism that aimed to defend Baconian and Cartesian inductive method. Kant's task was to explain how objects appeared to human awareness despite such scepticism about their true nature, and how human action was still possible in the world. The response that he offered, which has propelled modern understanding along its peculiar path to the present, was, as Leech argues, to render objects in an 'appearantist' space that the mind enlivened by virtue of its own 'transcendental' activity. The mind thus seemingly worked in ways analogous to the artist, superimposing the form and what could be known about the content of objects (but not their substance) onto a space-time 'canvas' whose existence had to be pre-supposed. The mind filled out this *a priori* canvas-like space-time entity instantaneously according to anthropological categories supplied directly by rational awareness. Since Kant was confident that rationality is an attribute of all human beings, he was also able to suggest that Hume's scepticism had rendered the problem of knowledge in too stark a way: although we may never know the thing in itself absolutely, we are equipped with a practical understanding about our relation to objects that works in fairly reliable ways – so long as we take care to separate practical reasoning from speculative reasoning and recognize certain limits obtaining between the two.

Kant was one of the first philosophers to offer a comprehensive theory of the mind. This fact alone argues for the importance of his philosophy when approaching consciousness. The turning point his philosophy marks for theories of consciousness consists precisely in the way he set out a role for a new kind of mental space that could be distinguished from the space of the outside world. The section in Kant's first *Critique* that grounds this theory onto an idea of inner space, the Transcendental Aesthetic, appears at the beginning of Kant's work, although Adorno is one Kant-interpreter who argues that it might better be considered at the end of his discussion, since this mind-led aesthetic initiates, but also re-claims, Kant's entire critical project. One might say that the Transcendental Aesthetic is the Platonic core of what seeks to grow an Aristotelian fruit, since the business of the *Critique* is ultimately to release human understanding from the tutelage of its own self-imposed prejudices in order to see clearly and in proper perspective, that is, ultimately, in a scientific way. This is the argument about the mature use of reason that Kant later unfolded in 1784 in his Enlightenment essay. The paradox of this release into the new science of mature rational understanding, from another angle, is that it avails itself of a conceit of art in order to ground its attenuated truth claims. The instinct to make science out of art in the Kantian argument is the insight on which many essays in the present volume are predicated.

Rose Woodcock, following J.J. Gibson, is one contributor who finds the space of pictures – the transcendental space of post-Cartesian philosophy in the manner of Kant – strictly speaking 'anomalous'. It is anomalous because it is something other than the mathematically governed geometrical space it was largely held to be in Kant's day, which is to say, it is not drawn solely in the manner of Alberti's orthogonals, and is not drawn as stereographic virtual reality requires.

This is because Alberti's perspectivalism made use of Euclidean assumptions about space that only allow for the three dimensions. The process of 'seeing into' this Euclidean space in fact involves a loss of the dynamic aspect of the subject – what James McArdle in his essay refers to as 'the force field in the landscape'. Euclidean space, even when drawn in solid form by art, that is, stereographically, imposes a necessary statement about its alienation from the purview of the subjective observer. For this reason, the mathematical perspectivalism of virtual reality actually appears antecedent to the space of pictures, because it has no self-observation, it contains no second-order statement about the conditions of its own constitution, no gap or interstice that is filled out by implication by subjects or creators who see and are aware of their seeing. Unlike Kant's transcendental aesthetic, which involves a second-order, 'Copernican' turn towards the subject in generating the conditions of possibility on which it depends (the categories might be consistent across the sum of human observers, but infinitely variable on an individual level within this broad consistency), virtual reality provides only for a single, Euclidean, absolute space. This space is something not experienced so much as partaken of. It moves us no closer to situating the dynamic consciousness that created it, and it knows of no self-relativizing will to its own creation.

This will to creation is only assayable once we estimate the role of the observer in the construction of space. McArdle's contribution, in engaging with the question of the animating presence of the observer, focuses not just on what is caught on the lens of the photographer as the object of depiction, but how the photographer's movement through space is the 'force field' that insinuates itself into the landscape and enters into a reciprocal relationship with it. The schematism of Euclidean geometry, for this reason, cannot account for the truth of the photographer's images; McArdle's metaphors are singularly non-Euclidean in their description of how objects are held together in the imaginative space of mental awareness. This does not leave Kantian thinking behind, but shapes it in the new direction of Merleau-Ponty's Kantianism, which argues for notions of incarnated awareness – dynamic utterances and self-created embodied visions – that speak of the 'poiesis' of the observer. Barry Empson's contribution offers similar insights into the situatedness of objects in musical space, arguing against the widely held view that music is nothing more than an unfolding of single musical events in time. The importance of the temporal dimension notwithstanding, Empson demonstrates with reference to Arnold Schoenberg that music has an undeniable relation to space, and in fact must presuppose such a space in a range of ways: in its construction of the tonal row, in its melodic lines that are only perceptible as Gestalt formations within an a priori space enlivened within the observer/auditor, even in its atonality. Thus, atonality describes the lack of orientation we feel in the musical space developed by the composer – the situation, we might say (adapting an image that Empson uses in reference to Schoenberg), where we are faced with a hat but are prevented from recognizing it as such no matter how we turn it.

The essays in the first part of the volume are brought to a conclusion by Louise Fairfax's essay on the unconscious in perception. Using a Kantian framework, Fairfax indicates what escapes direct notice in the schematic operation of rational perception. This is the problem of intentionality first developed into an explicit theory by Brentano and subsequently foregrounded in the phenomenology of Husserl and Merleau-Ponty: every perception has its own intentional aspect, and intention itself – the line of anticipation that forms the 'preobjective' patterning underlying observation – is equally structured by what eludes conscious intention. One might therefore say, with James Elkins, that every selection that takes place when we see into space, every intentional aspect that makes up the Gestalt formation that engenders vision, is also brought to life by what is not consciously intended – be it libidinous desire, a set of submerged preferences, an unacknowledged or unknown prejudice. These are forms of embodiment that drive our intentions just as much as our explicit intentionality, as Freud identified and encapsulated in a comprehensive theory. Something of this problem was already suggested a hundred years before Freud in Friedrich Jacobi's criticisms of Kant's first *Critique*. That Kant had failed to take account of the unspoken voluntary activity that underlies, and therefore undermines, the pretensions of our rationality, Jacobi held, suggested that critical philosophy offered no final alternative to religion, whose truths rescued the subject into the transcendent space of identity with the eternal. Our later view might now be, as Fairfax advocates, perhaps in consonance with the philosopher John Searle,[3] that we accept that all vision has an aspectual shape, and that 'consciousness does not represent the totality of experienced impressions'. It is not to Jacobi, but to Kant's compatriot Goethe, that Fairfax points in illustrating how we might better appreciate the reciprocal arrangements between observer and observed that apply when we see into space. Fairfax's allusions to Goethe indicate how the mental paradigm of consciousness was already shifting at the end of the eighteenth century toward a new understanding of embodied transcendental awareness, although we must also say, in view of Goethe's attempts to construct a dialogue on this point with Kant and Kant's failure to heed them that Kant himself had no appreciation of what Goethe was adding to his own account of this new transcendentalism.

The process of arriving at a new conception of mental space – something quite dissimilar from the space of the outside world in which human bodies are held – is also an account of the arrival of modern understanding, of what we refer to as modernity. In many ways, as the defenders of postmodernism would doubtless claim, it is still completely to arrive yet. Nonetheless, it seems fair to say that the process by which the patterns of Euclidean mathematics were revealed as nothing but the first layer, or a first attempt at layering, the mental space of the transcendental, was a business that took up the entire nineteenth century. The alternatives to a mathematically construed space were certainly announced early enough – one might look upon early German Romanticism, which inserted the supernatural into the natural and the everyday according to a literary-philosophical programme, as the first intimation of something different – but our literary and artistic chronology reveals that the attempt to add a non-Euclidean dimension to the Euclidean mathematical real was swamped by the arrival of a new realism, which only argued more insistently for the pre-eminence of the object in space over all kinds of subjectively intimated awareness. The early theories of the sublime, notably of Kant himself in the third *Critique*, also seem directed at

[3] Cf. Searle 1992: 131: 'Seeing an object from a point of view, for example, is seeing it under certain aspects and not others. In this sense, all seeing is "seeing as". And what goes for seeing goes for all forms of intentionality, conscious and unconscious. All representations represent their objects, or other conditions of satisfaction, under aspects. Every intentional state has what I call *an aspectual shape*' (original emphasis).

marking off the intentional, striated space of the conscious transcendental from the incursions, let us say, of the 'unstriated', and Kant's references to the 'mathematical sublime' in the *Critique of Judgement*, where limits are imposed even on the plane of abstract mathematical reasoning, perhaps indicate how great his intuitions were in this regard,[4] and how serious he thought the challenges to the mathematically calibrated space of the new science might prove to be. In art these challenges were increasingly announced in programmatic terms by the pre-Raphaelites and the Impressionists, but it wasn't until the late nineteenth and early twentieth century, ironically on the back of a revival of Kantianism in Germany and Austria, that a fundamental change to notions of space was clearly registered.

Henri Lefebvre, linking this change to the emergence of a new movement in art and letters in his pioneering work *La production de l'espace* (1974: 25), put it like this:

> The fact is that around 1910 a certain space was shattered. It was the space of common sense, of knowledge (savoir), of social practice, of political power, a space hitherto enshrined in everyday discourse, just as in abstract thought, as the environment of and channel for communications; the space, too, of classical perspective and geometry, developed from the Renaissance onwards on the basis of the Greek tradition (Euclid, logic) and bodied forth in Western art and philosophy, as in the form of the city and town. Such were the shocks and onslaughts suffered by this space that today it retains but a feeble pedagogical reality, and then only with great difficulty, within a conservative educational system. Euclidean and perspectivist space have disappeared as systems of reference, along with other former 'commonplaces' such as the town, history, paternity, the tonal system in music, traditional morality, and so forth.

Paul Monaghan's contribution, which links this new type of space to the appearance of Expressionism and thereby confirms Lefebvre's periodization, shows how Expressionists in the early twentieth-century felt themselves obliged to break with Symbolist assumptions about the 'theatre of the mind', for the mind alone could not suggest the complexity of the body's relation to outside space, and therefore could not render its 'anomalous' nature adequately. This was already a decisive step beyond the striated space of mathematics, which had

[4] As Heisenberg noted, Kant's notion of the 'thing-in-itself' is a 'mathematical structure' which, however, is not deduced directly from human experience (Heisenberg 1959: 83).

imposed a logic upon vision by a process of analogy with what was known about the law-bounded constitution of outer nature. Despite Helmholtz's commanding work on vision, which drew attention for the first time to the pre-rational, neurological patterning of what we see, the nineteenth-century did not break with a mathematically conceived nature seemingly moving in accordance with, and unfolding, an innate lawfulness, as supposed by Newton. Views of the natural world had yet to be confronted with the 'quantum' limitations that attend the perspective of the observer and problematize the observer's role in fashioning 'objective' truths. Although the early quantum theorists like Max Planck, Max Born and, later, Werner Heisenberg, were to break with this Newtonian foundation in fundamental ways,[5] the natural sciences in the early twentieth-century still lagged behind the insights of artists, who were now, as Monaghan's article describes, experimenting with techniques such as the use of stage stairs to indicate how the body might be conceived to enter and traverse spaces of transcendence in the theatre.

In the second set of essays, which is introduced by Monaghan's paper, art's imaginative construction of mental space thus becomes a theme. One of the first writers to view the library as a space of consciousness, with its stores of information amid labyrinthine corridors and passages, to be entered and exited more in accordance with Ariadne's thread than any predictive rationalism, was of course Jorge Luis Borges. Ewen Jarvis's article, situating itself in this Borgesian tradition, views Murakami, Eco and Canetti as 'cartographers of consciousness' whose perambulations in the space of the library reveal more completely than any realism the nature of our embodied awareness, with its dual attachment to the real and the unreal (Bachelard). The failure of realism to exhaust what can be depicted of the real perhaps comes about because realism is not focused on what Kim Roberts calls the 'intervals' between objects, and can therefore not account for the way interstitial moments not provided for under the terms of realism's 'real' take root in the consciousness of the

[5] As Born, for example, noted: 'Newton's time variable *t* is obviously an idealization abstracted from simple mechanical models and astronomical observations, fitting well into celestial motion, but not into ordinary experience' (Born 1949: 16). In 1933 Planck spoke of the difference between an idea of 'local force' defended by Newton and an 'integral force' of 'wave-mechanics' which in Newtonian thinking is not accounted for (Planck 1933: 63).

reader – a reader who, notwithstanding any pretended completeness of the artwork, is called upon to engage with the deficit that the intervals necessarily mark out. Roberts makes a suggestion about how we might imagine the interstices opened up by 'the bridge, the mirror, the labyrinth' in Calvino's postmodern fiction (where 'postmodern' as an epithet involves an open acknowledgement of incompleteness): these intervals are 'unheimlich', uncanny, they are gaps, as originally postulated by Fichte, between the 'I'-constructions of the subject and the 'not-I', where the world in which the subject moves but with which it reaches no identity cannot but remain, finally, a puzzle, an only partly experienced conundrum, as mysterious and inexhaustible as the early Romantics always took it to be.

Yet what is it of the world that the artwork has the pretense to reveal, or, put another way, what is it that observers experience when they enter into the space of the artwork? This question would appear to assume importance if we ascribe value, as the essays in this volume do, to the role of the observer in structuring the shared space that links art to its observation. If nothing in art is possible without the 'poiesis' of the observer, whose compact with the artwork creates art in ways first proclaimed, if somewhat rhetorically, by Roland Barthes's reference to the 'death of the author', can we approach this 'poiesis' of the observer in more than general ways, and can we describe or typify the experience that enlivens the artwork in the mind of the perceiver? This is the starting point of R.A. Goodrich's essay, which both brings the second group of essays to conclusion and establishes some parameters for the last sequence of questions posed in this volume. Using Anselm Kiefer's dramatic 'scorched earth' landscapes as a starting point, Goodrich moves from questions of representation, which would appear mainly to concern the activity of the artist, to notions of 'reenactment, acquaintance, or empathy', which bear upon the reciprocal space in which both artist and viewer are participants. The task of the viewer as participant is then described, following R.G. Collingwood, as the effort 'to reach, by thinking, the solution of certain problems arising out of the situation in which artists find themselves'. In extending from Collingwood's insight, Goodrich suggests that this engagement with the problems of the artist cannot be approached as a cognitive proposition alone, but ultimately arises as a question of empathy, as an involvement of the viewer's emotions in the problem scenario that the artwork discusses but does not, and perhaps cannot, finally resolve.

The essays in the third and last section of the volume flow from the problem of experiencing the artwork identified in Goodrich's approach. The tacit assumption of the entire volume – that consciousness occurs in a mental space that is both cognitive and emotional (even if we allow that these two terms could be separated), both free floating and yet time-bound and situated, both inward-facing to the habits of pure thought yet also compromised by the circumstances of the corporeal self – is made explicit in this section when embodied consciousness, in the guise of memory, is probed from a number of standpoints. As Dirk de Bruyn shows, even where artists examine their own art the situatedness of the artist and the artwork is a key consideration, suggesting in de Bruyn's case the certainty of the pain of a trauma partly acknowledged, and partly denied, and the givenness of 'fractured urban memories'. Goodrich had referred to Anselm Kiefer's statement that the artwork, in its memory-work, more closely approximates a 'burning' – what Thomas Bernhard in his novel *Auslöschung. Ein Zerfall* (1986) had elevated to the status of a programmatic 'annihilation' – than any transcendent construction distilling knowledge or final truth. The burning of which Kiefer speaks is perhaps another reference to 'the situation in which artists find themselves' (Collingwood), a situatedness, as de Bruyn illustrates, where the artist is forced to confront the psychic disturbances that register in memory in the post-Holocaust twentieth century even at the relative distance of Australia. From this angle, the difficulty of writing 'poetry after Auschwitz' (Adorno 1981) and facing a very real collective trauma in the second half of the twentieth century becomes a question for consciousness as much as it is a question for art.

Connecting with these insights, Khadija Carroll illustrates how vision is cast within a historical context that conditions what is witnessed and how it is witnessed. Where de Bruyn had discovered a gesture of erasure of the past in avantgarde 'direct theory', Carroll shows more generally how each act of naming of an object is not neutral, but subsumes the thing within a schema that itself is only decipherable if historical paradigms of understanding are first applied. Her interest is to highlight the historical nature of the givenness of objects in space and the necessary limitations of the vocabularies that are developed to deal with them. Thus she shows how the explorers on Cook's voyages to the south seas in the eighteenth century distinguished between the 'curious' – those objects that were classifiable according to taxonomies current in Cook's day and that frequently ended up in museum display cabinets – and the objects that were received with wonder, where 'wonder maintains an unaccountable remainder'. Eighteenth century 'curiosity' in her analysis is revealed as a control reflex governing an encounter with the incommensurable 'where a mutual unintelligibility threatens scientific order'.

If we follow Carroll's insights and assume, first, the absence of any neutral space of consciousness, and, secondly, that we are always already enveloped in a 'preobjective' space (Merleau-Ponty) that drives the circumstances of our own perception, how is pain issuing from the circumstances of this preobjective patterning ever to be confronted, much less ameliorated or overcome? Uli Krahn suggests a novel engagement with this problem, which in her analysis is also the problem of the situation of the artist, by 'changing the space of oneself' and thinking 'myself as statistic'. While this migration into a new space for conscious awareness adds the problem of migration, of geographical relocation to a circumstance of new embodiment, and is thus only another cross-layering in the dimensionality of the spatial, she nevertheless highlights the benefits that accrue when migration is used 'as a means of breaking out of the confining cycles of trauma'. What she has in mind connects with Carroll's ideas of taxonomic understanding, 'where characteristics are combined differently from the way they functioned in a previous self' according to the principles of 'recombinant hybridity'. In the new mental space the self inhabits, the part is loosed from the whole and a new Gestalt imaginatively remade. Using her own diary art as an example, Krahn engages with the problem of how stable identity formations known to govern the autopoietic self also render the self immobile under conditions of traumatic suffering. As she argues, complexity theory might indicate a way forward when one is able to conceive oneself 'as statistic': even minimal shifts over the passage of time might – as it were, probabilistically – effect a change in those identity formations that had been implicated in the construction of identity for the acute sufferer.

The movement from pure consciousness to embodied self set out in the trajectory of this volume is completed by Ann McCulloch's consideration of the problem of art and pain – a problem, as Adorno pointed out, that informs the reception and production of art in the post-Holocaust age, but a problem, too, that perhaps relates to the uniquely sensed time of the subject. Nietzsche's various criticisms of Kantian theory had run together on the question of embodied understanding: Nietzsche held that no practical utility for human beings flowed from the notion of the thing in itself, whatever the philosophical advantage in introducing it, and such a distinction had only led philosophers to embark on a fruitless search to recover it (Schopenhauer, for example, had equated the will with the 'Ding an sich' in his philosophy). One might view Nietzsche's genealogies – his narratives about the way we have acquired our values, and his view about the nihilism in religion which exercised such a profound influence on Freud – as attempts to make up the deficit left over in understanding after the incursions of pure thought of Kantian provenance. The rise of scientific reason, as he pointed out in *The Birth of Tragedy*, is scarcely a value-neutral affair, and obscures the shift that took place in the art of antiquity from a Dionysian pessimism that celebrated the senses to the Socratic sobriety and optimism that issued in rational understanding in later antiquity. It is only because of this shift, a change of emphasis willingly instigated and undergone by human beings, he further suggests, that we could have put science before art in the first place. It is out of this direct context that McCulloch draws her insights. Following Nietzsche's notion of the priority of art for mental life – that it is the narrative of the lived self that we compose creatively on the plane of our conscious life – McCulloch explores how our own art, our own creation of self, is shaped by the doctrine of eternal recurrence. In her reading of Nietzsche, eternal recurrence inhabits the space of the imagination as a mental predisposition that pre-exists our time-bound experience of life. One might say in the manner of systems theorists such as Niklas Luhmann or neurobiologists such as Maturana and Varela that eternal recurrence is the habit of repetition that engenders our 'autopoiesis' as a condition of the 'realization of the living' (Maturana/Varela). Without it we could have no identity, no capacity to realize ourselves as something separate from material life; without it there could be no statistical psychology of self. As McCulloch argues, this is not to suggest a 'mendacious idealism' (Nietzsche) that enjoins us to accept circumstances that we are unable to change. Rather it is the insight that gives us back our self in the midst of the pain of life with the recognition that 'there is pain that will not be obliterated'. The curve in McCulloch's argument therefore leads us away from the artificial constructions we might otherwise follow in the name of pure consciousness towards an Oedipal self that is necessarily blind in the face of the truths of disembodied rational understanding and must, on a psychological level at least, be concerned to break with them if our emotional life is to be fulfilled. While this may aim to re-animate, or, from another angle, fails to fore-stall, a new Dionysian pessimism, it also represents a platform on which the human condition might be accepted for its proclivities and chronic attachments no less than for the possibility of change and the achievements of mature vision, and thereby be affirmed.

I would like to thank the University of Otago, which contributed financial assistance during the first stage of this project. I also express thanks to the contributors, whose energy and diligence made working on this volume a joyful experience. Finally and most importantly, I thank Barry Empson for his invaluable support and assistance – always wisely and cheerfully given – in the preparation of this volume.

*References*

Adorno, Theodor W. 'An Essay on Cultural Criticism and Society.' T.A.: *Prisms*. Transl. Samuel and Shierry Weber. Cambridge, Mass.: MIT Press, 1981 (1949).

—. *Kant's Critique of Pure Reason*. Edited Rolf Tiedemann. Transl. Rodney Livingstone. Stanford, Calif.: Stanford University Press, 2001.

Alberti, Leon Battista. *On Painting*. Transl. with an introduction and notes by John R. Spencer. New Haven: Yale University Press, 1966 (1435/36).

Bachelard, Gaston. *The Poetics of Space*. Boston: Beacon Press, 1969.

Bacon, Francis. *The New Organon*. Edited Lisa Jardine, Michael Silverthorne. Cambridge, etc.: Cambridge University Press, 2000 (1620).

Barthes, Roland. *Image-Music-Text*. Transl. Stephen Heath. London: Fontana, 1977.

Benjamin, Walter. *Das Passagenwerk: Der Flaneur*. Edited Rolf Tiedemann. Frankfurt am Main: Suhrkamp, 1983.

Bernhard, Thomas. *Auslöschung. Ein Zerfall*. Frankfurt am Main: Suhrkamp, 1986.

Borges, Jorges Luis. *Labyrinths: Selected Stories and Other Writings*. Edited Donald A. Yates and James E. Irby. New York: Modern Library, 1983.

Born, Max. *Natural Philosophy of Cause and Chance*. The Waynflete Lectures. Oxford: Clarendon Press, 1949.

Brentano, Franz Clemens. *Psychology from an Empirical Standpoint*. Transl. Antos C. Rancurello, D.B. Terrell and Linda L. McAlister. London, etc.: Routledge, 1995 (1874).

Collingwood, R.G. *The Principles of Art*. Oxford: Clarendon, 1938.

Descartes, Rene. *Discourse on Method*. Introduction Ian Maclean. Oxford, etc.: Oxford University Press, 2006 (1637).

Elkins, James E. *The Object Stares Back: On the Nature of Seeing*. New York: Simon & Schuster, 1996.

Gibson, James Jerome. *Ecological Approach to Visual Perception*. Boston: Houghton Mifflin, 1979.

—. *The Perception of the Visual World*. Boston: H. Mifflin, 1950.

Heidegger, Martin. 'Die Zeit des Weltbildes.' M.H.: *Holzwege*. Frankfurt am Main: V. Klostermann, 1963 (1950).

Heisenberg, Werner. *Physics and Philosophy. The Revolution in Modern Science*. London: George Allen & Unwin, 1959.

Helmholtz, Hermann von. *Handbuch der physiologischen Optik*. 2 vols. Leipzig: Voss, 1867 (1856–67).

Hume, David. *A Treatise on Human Nature*. Edited Ernest C. Mossner. Harmondsworth: Penguin, 1969 (1739–40).

Husserl, Edmund. *Logische Untersuchungen*. 2 vols. Halle: M. Niemeyer, 1913.

Jacobi, Friedrich Heinrich. *Über das Unternehmen des Kritizismus, die Vernunft zu Verstand zu bringen*. F.H. Jacobi: *Werke* III. Edited Friedrich Roth and Friedrich Köppen, Darmstadt: Wissenschaftliche Buchgesellschaft, 1980 (1802).

Lakoff, George and Mark Johnson. *Philosophy in the Flesh: The Embodied Mind and its Challenge to Western Thought*. New York: Basic Books, 1999.

Lefebvre, Henri. *The Production of Space*. Transl. Donald Nicholson-Smith. Oxford, etc.: Blackwell, 1991 (1974).

Luhmann, Niklas. *Essays on Self-Reference*. New York: Columbia University Press, 1990.
—. *Observations on Modernity*. Transl. William Whobrey. Stanford: Stanford University Press, 1998.
Kant, Immanuel. 'On the Question: What is Enlightenment?' *What is Enlightenment? Eighteenth-Century Answers and Twentieth-Century Questions.* Edited James Schmidt. Berkeley, etc.: University of California Press, 1996, 58–64 (1784).
—. *Kritik der reinen Vernunft*. Hamburg: Fritz Meiner, 1956 (1781/1787).
—. *Kritik der Urteilskraft*. Stuttgart: Philipp Reclam, 1956 (1790).
Merleau-Ponty, Maurice. *Phenomenology of Perception*. Transl. Colin Smith. London: Routledge & Kegan Paul, 1962 (1945).
Maturana, Humberto, and Francisco J. Varela. *Autopoiesis and Cognition. The Realization of the Living*. Dordrecht, Holland: D. Reidel Pub. Co., 1980.
Nietzsche, Friedrich. *The Birth of Tragedy and Other Writings*. Edited Raymond Geuss and Ronald Speirs. Transl. Ronald Speirs. Cambridge, U.K. etc.: Cambridge University Press, 1999 (1872).
Panofsky, Erwin. *Perspective as Symbolic Form*. New York: Zone Books, 1991 (1927).
Planck, Max. *Where is Science Going?* Woking: Unwin Brothers, 1933.
Ramachandran, Vilayunar. *The Emerging Mind: The Reith Lectures 2003*. London: Profile, 2003.
Schopenhauer, Arthur. *Die Welt als Wille und Vorstellung*. 2 vols. Zürich: Haffmanns Verlag, 1991 (1818–19; 1844).
Searle, John. *The Rediscovery of the Mind*. Cambridge, Mass.: MIT Press, 1992.
Snow. C. P. *The Two Cultures and the Scientific Revolution*. New York: Cambridge University Press, 1959.
Varela, Francisco J. and Evan Thompson, Eleanor Rosch. *Embodied Mind: Cognitive Science and Human Experience*. Cambridge, Mass.: MIT Press, 1991.
Wilson, Fred. *Logic and Methodology in Early Modern Thought: Seven Studies*. Toronto, etc.: University of Toronto Press, 1999.
Wittgenstein, Ludwig. *Tractatus Logico-Philosophicus*. Introduction Bertrand Russell. London, etc.: Routledge and Kegan Paul, 1985 (1922).

# PERCEPTION, PERSPECTIVE, REPRESENTATION

# The Painting of Philosophy: Space, Perspectivalism, Representability and Consciousness

Peter Leech

Perspective is a branch of philosophy.
Michele Savonarola c1440

Immanuel Kant was no connoisseur of the arts. A selective inventory of aesthetic examples from his *Kritik der Urteilskraft* is really rather disheartening: foliage on wallpaper, a mawkish poem by King Frederick of Prussia, 'some' crustacea, and 'New Zealanders with their tattooing' (Kant 1790: 73).[1] Music in particular displeases Kant: it 'has a certain lack of urbanity about it', he asserts, and adds – though it seems uncommonly early for the ghettoes of Königsberg to be infested with blasters – that it 'scatters its influence abroad to an uncalled-for extent (through the neighbourhood) [...] and becomes obtrusive' (Kant 1790: 196).

It is, then, with some apprehension that one turns to Kant's consideration of painting. Mercifully, perhaps, he offers no example. There is in fact a foreboding silence from his biographers about whether Kant had actually *seen* one, though I am sure he had – if only on the grounds that he boldly contends that colours in a painting are merely 'charming' and therefore *not* grounds to make it 'really worth looking at' (Kant 1790: 67). But now comes the surprise in what remains the foundational study of aesthetic experience. For in a late section of Kant's Third Critique, we find that painting actually ranks very high in his estimation: '[a]mong the formative arts', he reports, 'I

[1] Citations of the *Kritik der Urteilskraft* 1790 are from Kant, *Critique of Judgement,* transl. J.C. Meredith, Part One, Critique of Aesthetic Judgement, with Meredith's pagination signalled in the text (see references at the end of this article).

would give the palm to *painting*'; and it is part of Kant's reasoning here that 'painting can penetrate much further into the region of ideas, and in conformity with them give a greater extension to the field of intuition than it is open to the others to do' (Kant 1790: 196; Kant's emphasis). Music at dinner may be, for Kant, 'an agreeable noise' intended to foster 'a genial' spirit of conversation (Kant 1790: 166). Yet painting, it seems, can actually *picture* philosophy. Hence my title, 'The Painting of Philosophy', and, in amplification, my epigraph from the Italian writer, Michele Savanorala – fully three centuries before Kant – in his declaration that '[p]erspective is a branch of philosophy.'[2] My purpose generally is to argue that the conception of perspectival space – first elaborated in Savonarola's *quattrocento* Italy – is indeed as much a philosophical as a pictorial idea, and that it has a profound bearing on what I shall call Kant's 'philosophical perspectivalism', particularly in his exploration of our consciousness of space in his First Critique – the magisterial *Kritik der reinen Vernunft* (1781/7).

In sequence, my argument follows this path. First, under the heading 'Philosophy and Painting', I want to consider a collusive moment of painting and philosophy – and a very specific metaphysical context for the discussion – which may be found in the work of another great philosopher, Bertrand Russell. And though Russell is not here openly concerned with space, there is a promising analogue in pictorially spatial terms. In 'A Transcendentally Aesthetic Perspective', I shall attempt to develop this analogue via the opening section of the First Critique (the 'Transcendental Aesthetic') in which Kant introduces his discussion of space. I propose that his thinking might be construed *both* as a form of philosophical perspectivalism *and* as an implied commentary on perspectival painting. Finally, in 'Philosophical Perspectivalism and the Representability of Consciousness', I shall try further to secure Savonarola's observation that '[p]erspective is a branch of philosophy' with broader reflections on non-Euclidean space, on non-perspectival painting and two piercing philosophical comments by Ludwig Wittgenstein on the nature of our visual consciousness.

[2] Michele Savonarola, *Libellus de magnificis ornamentis regie civitatis Padue* 1440c. Cited in Lightbown 1986: 22.

*Philosophy and Painting*

There is a remarkable passage of thought about painting which introduces Bertrand Russell's astute little book, *The Problems of Philosophy* (1912). 'The painter', Russell suggests, 'has to unlearn the habit of thinking that things seem to have the colour which common sense has they "really" have, and to learn the habit of seeing things as they appear' (Russell 1967: 2). To this sentence Russell immediately adds: 'Here we have already the beginning of one of the distinctions that cause most trouble in philosophy – the distinction between "appearance" and "reality"' (1967: 2).

That Russell should focus on the painter's *colour* seems to me interesting for two reasons. The first is historical. Writing in 1912, the kind of contemporary painting with which Russell was likely most familiar was Impressionism, whose precepts – more or less dominant since the first Impressionist exhibition in 1874 – had only just come under attack from the Cubists. And what Russell has to say, construed as a comment on Impressionism, might well – forty years earlier – have silenced the idiot-critics who initially condemned Impressionism. 'They put *green* in their shadows!', sneered one of them, Marc de Montifaud, never apparently having observed the appearance of, say, an expanse of partly-shaded meadow or river-bank.[3]

My second reason for finding this interesting: Russell *does* choose a philosophically uncontroversial and very *British* – British Empiricist – example of what the painter might do. Along with John Locke, for instance, Russell would certainly have taken colour to be a *distinctively* 'secondary', not a 'primary' quality: a function, that is, of our *sensations* of an object rather than a property of the object itself. It is difficult here to imagine Russell's thinking of *spatial* consciousness in such terms; and of course, for Locke, spatial qualities such as extension and figure are 'primary' not 'secondary'. In fact I would suggest here that for Locke – perhaps for all empiricist philosophers

[3] Marc de Montifaud [pseudonym of Marie-Amélie Chartroule de Montifaud], 'Exhibition du boulevard des Capucins', *L'Artiste* (1 May 1874), cf. Adhemar and Gache 1974: 256 (author's translation).

(including Russell) – spatial qualities *alone* are definitive of properties of an object or body.[4]

Kant would certainly not have approved of Russell's example of the painter's colours, even had he had the opportunity to confront an Impressionist painting, and this is *not* just because, as I earlier noted, he found colours merely 'charming' and therefore not grounds to render a painting 'really worth looking at'. Indeed – though this remark stands in need of a contextualisation which I shall later offer in terms of the Kantian concept of 'appearance' in the First Critique – Kant observes that: 'Colours are not properties of the bodies to the intuition of which they are attached but only only modifications of the sense of sight, which is affected in a certain manner by light [...] changes in the subject, changes which may, indeed, be different for different men' (Kant 1781/7: 73).[5] Still, there *is* a type of painting which one can say with certainty – despite biographical silence on the matter – that Kant *did* confront. This is nothing so specific as a *movement* of painting such as Impressionism, but rather a representational *form* of painting: namely, perspectival painting, here emblematically represented by Luciano Laurana's depiction of *La Città Ideale (The Ideal City),* c1470.[6]

Yet why – it will be asked – can I assert this with such confidence? Well, the answer is that from at least the mid-fifteenth-century until at least the early twentieth-century – including Impressionism – *all* (competent) paintings are perspectival in representational form. Kant *could not have viewed* non-perspectival paintings, unless very bad attempts at perspective, or if – as it is inconceivable to suppose – he had some interest in, or access to, pre-perspectival paintings from the fourteenth century or earlier. And it is precisely around this point that the *quattrocento* origination of perspectival theory, and the space

---

[4] For Locke's original distinction between 'primary' and 'secondary' qualities, see Locke 1964: 112–119 and 135–143.

[5] Citations of the *Kritik der reinen Vernunft* (either the first edition of 1781, or second edition of 1787) are from *Immanuel Kant's Critique of Pure Reason,* transl. Norman Kemp Smith, with Kemp Smith's pagination signalled in the text (see References).

[6] *La Città Ideale* is one of three paintings with the same theme associated with a principal centre of perspectival studies at the magnificent ducal palace of Federico da Montefeltro (Duke of Urbino 1444–72). This particular example, formerly attributed to Piero della Francesca but now to Luciano Laurana – chief architect of the ducal palace – remains in the palace, now the Galleria Nazionale delle Marche, Urbino.

which it portends, lends itself to Savonarola's observation that perspective is a branch of philosophy.

It is commonplace in the history of art to construe that origination in Leon Battista Alberti's *De Pictura* simply as a set of geometric rules – a *method* – for the construction of a tiled pavement in perspective: the *costruzione legittima*, as it is usually called. But it is, I suggest, far *more* than a pictorial method. Alberti himself makes modest apology for the 'obscurity and difficulty' of his mathematics, but – interestingly – none for the covert philosophy which underlies it. Indeed, as he puts it, '[t]o intelligent minds that are well disposed to painting, those things are simple and splendid, however presented, which are disagreeable to gross intellects' (Alberti 1435: 59).[7]

However, there are also at least four genuinely philosophical points Alberti makes which might be brought together. First, the *costruzione legittima* itself concludes thus: 'We will now go on to instruct the painter how he can represent with his hand what he has understood with his mind' (Alberti 1435: 59). Alberti, that is, is far from thinking that what he has offered is a set of *manual* instructions. Second, he says (rather awkwardly, it's true), 'no learned person will deny that no objects in a painting can appear like real objects unless they stand to each other in a determined relation' (Alberti 1435: 57). Third – in respect of the 'determined relation' he has in mind between the appearance and reality of objects – Alberti offers a direct invocation of philosophy in citing Protagoras's conviction that 'man is the scale and the measure of all things' (Alberti 1435: 53). Fourth – in what might be mistaken as mere general intellectual advice, but which I suggest has both a philosophical *and* a pictorial dimension, Alberti proposes that '[i]t is useless to draw the bow, unless you have a target to aim the arrow at' (Alberti 1435: 59). And in respect of this fourth remark, let me point to a pictorialization of this metaphor in a startling detail from Andrea Mantegna's now ruined 1448 fresco *The Martyrdom of St Christopher*, in which we find that the target *is* a human eye.[8] As

---

[7] Citations of the *De Pictura* (1435) are from Leon Battista Alberti, *On Painting and On Sculpture*, transl. Cecil Grayson, with Grayson's pagination signalled in the text (see References).

[8] Mantegna's work (in the Ovetari Chapel, Padua) depicts the giant St Christopher's punishment for refusing to make sacrifice to the pagan gods. In the first (left-side) scene King Dagnus of Lycia orders his archers to kill Christopher, but their arrows

Michael Kubovy has put it: 'I see the arrow in the eye as a metaphor for the art of perspective' (Kubovy 1986: 1).[9]

But just before returning to Kant, let me sum up what I take to be Alberti's philosophical propositions in his writing about perspectival space in painting. Such a pictorial space is *not* simply an artifice of painting – a convention constrained by the attempt to represent three dimensions in two. It is how space actually *appears* (visually) to the human beings who scale and measure it. This is, indeed, how we *do* – or *should* – understand it 'with our minds' (philosophically-sensitive minds, that is). The target of pictorially-perspectival space *is* the eye and how space appears to the eye, notwithstanding the ubiquitous error of supposing a perspectivally-correct painting to be a 'real-ist' painting, rather than – as I always want to insist – an *'appearant-ist'* painting. The matter really is, as my earlier quotation of Russell suggested, an example of 'the beginnning of one of the distinctions that cause most trouble in philosophy'.

### *A Transcendentally Aesthetic Perspective*

Where – many would say – the Russellian trouble of a purely philosophical kind really does emerge is in the opening discussion of space in the 'Transcendental Aesthetic' of Kant's *Kritik der reinen Vernunft*. For it is here that Kant first establishes his notorious distinction between phenomena and noumena – between things as we experience them and things as they are in themselves. It is perilous to attempt to negotiate here a comprehensible path through the complexities of the Transcendental Aesthetic in a manner which would neither outrage professional philosophers (Kantians or not) nor simply mystify non-philosophers. Let me try to reduce the risk by coming to it indirectly. As I earlier hinted, in noting the date of

---

remain suspended in mid-air. When Dagnus mocks Christopher, believing him dead, an arrow falls from the air and pierces his eye, rendering him blind. In the second (right-side) scene, Dagnus's blindness is cured by smearing his eye with Christopher's blood, and the beheaded saint is dragged away. For details of the fresco, see Ronald Lightbown 1986: 387–400.

[9] There are in fact very good reasons for Kubovy's belief. As I have argued elsewhere, Mantegna appears to have read Alberti's *De Pictura* extremely closely, and this is not the only instance of a pictorialisation of a passage from the text. For details of the relations between Mantegna and Alberti, see 'Mantegna e Alberti' in Muraro 1965: 103–132.

Russell's remark about what the painter does, Impressionism was in 1912 only just coming under attack from the Cubists and their theoretical advocates. And the point of attack *was* essentially philosophical. As Juan Gris – the most philosophical of the Cubist painters – was to put it: 'By way of natural reaction against the fugitive elements employed by the Impressionists, painters felt the need to discover less unstable elements in the objects to be depicted.'[10] It is plain that the 'fugitive elements' to which Gris refers *are* (visual) sensations. Indeed, the uniquely sharp critic, Jules-Antoine Castagnary, had actually said of the painters in 1874 that '[t]hey are *impressionists* in the sense that they render not the landscape but the sensations produced by the landscape' – a claim which might be said to make the painters 'Impressionists' virtually in David Hume's sense of the term.[11] Further, it may not be too fanciful to suppose that 'the *less* unstable elements in the objects to be depicted' are something like Locke's 'primary qualities', and I am especially inclined to suppose this on the grounds that a particular target of Cubist contempt *was* the perspectival representation of space: 'that miserable, tricky perspective [...] that infallible device for making all things shrink', as Apollinaire once put it (Apollinaire 1913: 66; author's translation). To condense a vastness of vituperation around this issue, let me invoke two exquisitely succinct comments. First, the astute theorist of Cubism, John Golding, makes this point, perfectly illustrated by Juan Gris's 1914 painting, *The Teacups*. 'Knowing that the opening of a cup is round, it is false to depict it simply as an ellipse' (Golding 1959: 33).[12]

---

[10] Juan Gris, in response to a questionnaire, 'Chez les Cubists' in *Bulletin de la Vie Artistique,* Vol. VI, No.1, Paris, January 1925. Cited in Kahnweiler 1947: 144–5 (Cooper's translation).

[11] 'Ils sont *impressionistes* en ce sens qu'ils rendent non le paysage, mais la sensation produite par le paysage': Jules-Antoine Castagnary, 'Exposition du boulevard des Capucins', *Le Siècle* (29 April 1874), in Riout 1989: 56–7 (Castagnary's emphasis; author's translation). It is an intriguing – and largely unexamined – feature of the history of philosophy that Hume wrote *A Treatise of Human Nature* (1739) while resident in France, and that it may have had more immediate impact on French than on British philosophy. For hints of this significance, see Schiff 1986: 61–89.

[12] It is specially interesting that in this passage Golding draws attention to the writings on Cubism by Olivier Hourcade and Hourcade's references to Kant.

Now 'false', here, is an interestingly muscular diagnosis, but it certainly captures the spirit of Picasso's bluntness in asserting that 'I paint objects as I *think* them, and not as I *see* them'.[13] Only it is *exactly* between Picasso's *thinking* of, and his *seeing* of, space that it seems to me Kant's Transcendental Aesthetic intervenes. For Kant, our experience of the world is given to us in what he calls 'intuitions', and this mode of experience 'in which we are affected by objects, is entitled *sensibility*' (Kant 1781/1787: 65). From this one might suppose that our consciousness of space and spatial properties is simply a function of the sensations we happen (empirically) to have. And indeed from this one might further deduce the prompt to Cubist agitation not only over perspective but about any painting which attempts to render mere sensations.

Kant, however, insists on a distinction between *two* kinds of 'intuition': what he calls on the one hand 'empirical intuition', and on the other 'pure intuition'. He explains this distinction via another: that is, between the '*matter*' (content) of appearance, and 'the *form* of appearance' (Kant 1781/1787: 65–66). Instances of the *matter* of appearance here are, say, colour-sensations which are, properly, for Kant (for example, in the passage I cited earlier) 'only modifications of the sense of sight, which is affected in a certain manner by light'. Yet as for the *form* of appearance, there are two kinds of 'intuitions' which are 'pure' (purely formal) and *a priori.* These are the pure intuitions of space and time, which are the subject of Kant's reflections in the Transcendental Aesthetic. Together (though I have nothing here to say of time) they constitute not the *content* of our sensibility, but rather the very *conditions* of it. That is, as Kant puts it, '[b]y means of outer sense, a property of our mind, we represent to ourselves objects as outside us, and all without exception in space. In space their shape, magnitude, and relation to one another are determined or determinable' (Kant 1781/1787: 67).

The question now is how Kant's arguments might be said to bear on perspectival representation. By way of answer, let me first return to the passage of Alberti, cited in the preceding section in which he

[13] Pablo Picasso, quoted in R. Gomez de la Serna in 'Completa y Veridica Historia de Picasso y el Cubismo' in *Revista del Occidente* (Madrid, July-August 1929) – cf. Golding 1959: 60 (author's emphasis).

claims that 'no learned person will deny that no objects in a painting can appear like real objects unless they stand to each other in a determined relation' (Alberti 1435: 57). Only *what* is here doing the spatially relational determining? For Kant, as I have just cited him to say, it is 'a property of our mind'. But for Alberti, the determining factor is 'nature herself'. In a curious conclusion to his *costruzione legittima*, Alberti seems to suppose that the representation of space in perspectival form is *just* empirically derived. As he puts it: 'This is why men depicted standing in the parallel [of a pavement] furthest away are a great deal smaller than those in the nearer ones – a phenomenon which is clearly demonstrated *by nature herself*, for in temples we see the heads of men walking about, moving at more or less the same height, while the feet of those further away may correspond to the knee-level of those in front' (Alberti 1435: 57; author's emphasis).

Yet this claim *cannot* be about 'nature herself'. Human figures do *not* diminish in size the further they are away: it is absurd to say so, and a special point of ridiculousness in thinking of perspectival paintings as 'real-ist' rather than – as I earlier suggested – 'appearant-ist'. Despite the oddity of his claim, however, perhaps Alberti might be defended in noting that he does, confusingly, seem also to be talking about mere human visual sensation ('we *see* the heads of men walking about'). But *neither* claim would do for Kant. Space, as a pure intuition – insofar as I have so far been able to explain his point – is *not* simply a property of the world ('in-itself' – independently, that is, of our experience of it). Yet *nor* is it simply a property of (visual) sensation (as, for example, colour is). And this, I suggest, is the Cubists' fundamental (philosophical) mistake. In eliminating, say, Impressionist visual 'appearances' it does not follow – at least in Kantian terms – that the painter consequently depicts 'things-in-themselves'.[14]

There are, however, two points which might be said to draw Alberti and Kant into a sort of complicity. The first concerns Alberti's

[14] Olivier Hourcade makes this mistake. In his article, 'La Tendance de la Peinture Contemporaine', in *La Revue de France et des Pays Français* (February 1912) he explicitly invokes Kant on behalf the Cubists by quoting Schopenhauer: 'Le plus grand service que Kant ait rendu c'est la distinction entre le phénomène et la chose en soi'; cf. Golding 1959: 33.

quotation from Protagoras which I noted earlier: namely, that 'man is the scale and the measure of all things'. For Kant makes an invitingly similar point in his Transcendental Aesthetic: namely, that '[i]t is, therefore, solely from the human standpoint that we can speak of space, of extended things, etc.' (Kant 1781/1787: 71). The second point concerns geometry, and a deep problem in the theory of perspective itself, of which I shall indicate no more here than that the theory itself has – from the start – hovered uncertainly between geometry (which one might be inclined to call an *a priori* science), and optics (which one might be inclined to call an *a posteriori,* or empirical science). For instance, in his seminal, if non-philosophical, book on perspectival theory, *The Poetics of Perspective,* James Elkins makes the point that 'optics and mathematics are unevenly distributed in proofs [of perspectival representation], and no one is quite sure how to handle them together. The choice between optics and mathematics – it continues to appear as a choice, even when the two are said to be strongly related – is never quite made' (Elkins 1994: 12).

It is certainly the case that the fundamental principles of Alberti's *costruzione legittima* are presented as a geometric proof. And this is even more emphatically the case with Piero della Francesca's proof of linear perspective in his *De Prospectiva Pingendi* (1470c). But how, then, are we to understand geometry *itself*? Interestingly, and unlike Alberti, Piero is far more inclined to suppose the *a priori*-ty of geometry in saying: 'I therefore assert the necessity of perspective which, as a true science, discerns every quantity proportionately and demonstrates their diminution and enlargement *by the power of lines*' (Francesca 1984: 129; author's translation and emphasis). Kant, in his Transcendental Aesthetic, is peculiarly alert to this kind of problem (if not specifically in terms of perspectival representation). 'Geometry', he says, 'is a science which determines the properties of space synthetically, and yet *a priori.* What, then, must be our representation of space, in order that such knowledge of it may be possible?' (Kant 1781/1787: 70). Kant's answer to his own question is that our representation of space 'must be found in us prior to any perception of an object, and must therefore be pure, not empirical intuition', and this, he adds, is 'the only explanation that makes intelligible the *possibility* of geometry, as a body of *a priori* synthetic knowledge' (Kant 1781/1787: 70–1).

By way of introducing my last section, 'Philosophical Perspectivalism and the Representability of Consciousness', there is a problem which will have driven certain kinds of philosophers to barely suppressed abuse. The problem is this: in talking intermittently about perspective in painting and then about Kant, it is likely to be objected that I have been talking about the *representability* of space, not about space itself. I immediately accept the point, both on behalf of perspectival painting *and* philosophy. Only, I would say, the entire *burden* of Kant's Transcendental Aesthetic is that we can say nothing about 'space-in-itself': we *can* only talk of its *representation*. And in defence of this claim I mean to move finally to some thoughts of Wittgenstein.

*Philosophical Perspectivalism and the Representability of Consciousness*

When Kant said, as already cited, '[i]t is, therefore, solely from the human standpoint that we can speak of space, of extended things, etc.' (Kant 1781/1987: 71), he maintains what I have called a form of 'philosophical perspectivalism': he would have liked, I fondly imagine, the detail I showed from Mantegna's *Martyrdom of St Christopher* with its proposed perspectival metaphor. In fact this might be said to be a very good reason for his high estimation of painting on the grounds that it 'can penetrate much further into the region of ideas, and in conformity with them give a greater extension to the field of intuition' (Kant 1790: 196). A mandarin response to this kind of 'humanistic' perspectivalism – certainly a response which Cubist commentators (though not the Cubists painters themselves, interestingly) were attracted to – is to suggest that there is, '*of course, now* (c1912)', an entirely different conception of space of which the original perspectival painters and theorists, and importantly Kant himself, were quite unaware: namely, non-Euclidean geometry.[15] To

[15] See for instance Albert Gleizes and Jean Metzinger, *Du Cubisme* (1912) and Guillaume Apollinaire, *Les Peintres Cubistes* (1913). Gleizes and Metzinger claimed that: 'If we wished to refer the space of these [Cubist] painters to geometry, we should have to refer it to the non-Euclidean scientists: we should have to study at some length certain theorems of Riemann's [the nineteenth-century mathematician]', cited in Chipp 1968: 212. Apollinaire, more excitably, claimed that: 'Today, scientists no longer limit themselves to the three dimensions of Euclid. The painters have been led

enter that domain in any detail is not sensible on this occasion, though I should note that the development of non-Euclidean geometry is held by some – notably Alan Musgrave – to refute Kant's conception of geometry as synthetic but *a priori*.[16] A couple of general remarks are appropriate.

The great French mathematician, Henri Poincaré, pointed out that 'experience brings us into contact only with representational space, which is a physical continuum, never with geometrical space, which is a mathematical continuum' (Poincaré 1958: 69). Poincaré, it is true, maintains that Euclidean principles (well, at least one: the parallel axiom) are refutable and have a logical alternative, but also that no one would ever actually adopt the non-Euclidean alternatives because – as Musgrave himself says in a comment on Poincaré – 'the resulting system of science will be too complicated' (Musgrave 1993: 234). Yet 'too complicated' for what purpose, I ask? I *suspect* that the point here is not simply a matter of scientific pragmatism, but rather that a non-Euclidean geometry of space is just not *visually representable* in any coherent or comprehensible way. It can be *described,* or perhaps better, *suggested* in principle, sure. But so too can an object which is simultaneously red and green all over. I just *have* so described and suggested one – except, of course, *no* such object could be pictured.

Consider, for example, a non-Euclidean 'hyperbolic space' which purports to refute Euclid's parallel axiom the essential burden of which is that there is just *one and only one* straight line parallel to a given straight line (that is, no matter how far extended, that straight line will never intersect the given straight line). But in a 'hyperbolic' finite and bounded space (finitely bound, for example by a circle) there are infinitely many chords of the circle which cannot intersect a given straight line enclosed within the circle, since *ex hypothesis* they cannot be extended beyond the finite, bounded space. According to the Euclidean principle, such chords will be 'parallel' to the enclosed

---

quite naturally [...] to preoccupy themselves with the new possibilities of spatial measurement which are designated by the term: *the fourth dimension*', cited in Chipp 1968: 223.

[16] See in particular Musgrave 1993: 232–5. I would like to thank my friends Alan Musgrave and David Ward (both of the Department of Philosophy, University of Otago) for many years of philosophical brawling over Kant: Alan for his irremediable Kantian antipathies, David for his equally irremediable Kantian sympathies.

straight line – yet none, if I can put it this way, point in the same direction. Such a space is indeed a 'hyperbole': for though it can be diagrammatically described or suggested, a world of which this might be true is indeterminably chaotic in visual terms. It is the visual chaos, I suggest, of Ambrogio Lorenzetti's pre-perspectival painting of 1340, *City by the Sea*, or of the Cubists' post-perspectival painting (for example, once more, Gris's *The Teacups*).

In fact this charge of 'chaos' *is*, surprisingly, made of Cubist paintings by one of the most critically astute and generally *supportive* commentators on the Cubists, Jacques Rivière, who, in an article of 1912 (with a section entitled 'Why perspective must be eliminated'), nevertheless concludes by pointing out that 'the elements in their [Cubists'] pictures relapse into anarchy and form a mad cacophony' (Rivière 1992: 186). What is at issue here is the *determinability* of a representation (precisely the point I have already emphasized in relation both to Alberti on perspectival painting and to Kant on what I am calling his 'perspectival philosophy'). To reduce massive matters to a tiny pinpoint, consider this. In emphasizing Poincaré's (and Musgrave's) point that non-Euclidean geometry is simply a logical, or conceptualizable, alternative to Euclid's, Morris Klein makes the claim that non-Euclidean axioms 'depend only on reasoning and not at all on accord with diagrams' and adds that 'the failure to correspond with visual sensations is irrelevant' (Klein 1954: 416).

My concluding question regarding space and its representability for consciousness, however, must be: *Is* this irrelevant? By way of an answer I want finally to turn to Wittgenstein.

In his typically gnomic manner, Wittgenstein once asked: 'What is the criterion of the visual experience? – The criterion? What do you suppose? The representation of "what is seen"' (Wittgenstein 1972: 198e). The phrase 'visual experience' here, might *seem* to give over to the anti-Kantian, non-Euclidean, philosopher of space all that he needs. But I think, actually, not. For immediately following this paragraph Wittgenstein adds this: 'The concept of a representation of what is seen, like that of a copy, is very elastic, and so *together with it* is the concept of what is seen. The two are intimately connected. (Which is *not* to say they are alike)' (Wittgenstein 1972: 198e). This is, *precisely* I think, a way of understanding Kant's 'Transcendental Aesthetic'. The concept of a (visual) representation of space, and the concept of space itself may indeed not be 'alike': quite clearly they are *not* in any perspectival painting, and that is the common error I referred to earlier in the common supposition that perspectival painting is some sort of 'realism', rather than my 'appearantism'. Yet the two *are* – I would claim, as would Kant, along with Wittgenstein – 'intimately connected'.

It is the intimacy of this connection for our consciousness between the concept of space and its visual representation in painting which has concerned me. Russell is right to have pointed to the 'trouble' for philosophy, but Wittgenstein points to the real expense of it all. In *Culture and Value* he has this to say: 'The human gaze has the power of conferring value on things; but it makes them cost more too' (Wittgenstein 1980: 1e). The palm, that is, which Kant thought to award to the (perspectival) painting of philosophy, is indeed a rich honour.

### *References*

Adhemar, Hélène and Sylvie Gache (eds). *Centenaire de l'impressionisme.* Paris: Grand Palais, 1974.

Apollinaire, Guillaume. *Les peintres cubistes: Meditations esthétiques.* Paris: Figuière, 1913.

Battista Alberti, Leon. *On Painting and On Sculpture*. Transl. Cecil Grayson. London: Phaidon, 1972.

Chipp, Herschel B. (ed.). *Theories of Mondern Art.* Berkeley: University of California Press, 1968.

Elkins, James. *The Poetics of Perspective.* Ithaca: Cornell University Press, 1994.

Francesca, Piero della. *De Prospectiva Pingendi.* Ed. G. Nicco-Fasola. Firenze: Casa Editrice le Lettere, 1984.

Golding, John. *Cubism: A History and An Analysis 1907–1914.* London: Faber and Faber, 1959.

Kahnweiler, D.-H. *Juan Gris: His Life and Work.* Transl. Douglas Cooper. London: Lund Humphries, 1947.

Kant, Immanuel. *Critique of Judgement.* Transl. J.C. Meredith. Oxford: Oxford University Press, 1952.

—. *Immanuel Kant's Critique of Pure Reason.* Transl. Norman Kemp Smith. London: Macmillan, 1968.

Klein, Morris. *Mathematics in Western Culture.* London: George Allen and Unwin Ltd, 1954.

Kubovy, Michael. *The Psychology of Perspective and Renaissance Art.* Cambridge: Cambridge University Press, 1986.

Lightbown, Ronald. *Mantegna.* Berkeley, and Los Angeles: University of California Press, 1986.

Locke, John. *An Essay Concerning Human Understanding.* (1690). Ed. A.D.Woozley. London: Collins, 1964.

Muraro, Michelangelo. *Arte, Pensiero e cultura a Mantova nel primo Rinascimento in rapporto con la Toscana e con il Veneto: Atti del VI Convegno internazionale di Studi sul Rinascimento.* Florence: Istituto Nazionale di Studi sul Rinascimento, 1965.

Musgrave, Alan. *Common Sense, Science and Scepticism.* Cambridge: Cambridge University Press, 1993.

Poincaré, Henri. *The Value of Science.* Transl. George Bruce Halsted. New York: Dover Publications, Inc., 1958.

Riout, Denys (ed.). *Les écrivains devant l'impressionisme.* Paris: Editions Macula, 1989.

Rivière, Jacques. 'Present Tendencies in Painting' (1912). Transl. Edward Fry. *Art in Theory 1900–1990: An Anthology of Changing Ideas.* Charles Harrison and Paul Wood (eds). Oxford: Blackwell, 1992. 183–7.

Russell, Bertrand. *The Problems of Philosophy.* (1912) Oxford: Oxford University Press, 1967.

Schiff, Richard. 'The End of Impressionism.' Charles S. Moffett, et al.: *The New Painting: Impressionism 1874–1886.* Geneva: Richard Burton SA, 1986. 61–89.

Wittgenstein, Ludwig. *Culture and Value.* Transl. Peter Winch. Chicago: University of Chicago Press, 1980.

—. *Philosophical Investigations.* Third Edition. Transl. G.E.M. Anscombe. Oxford: Basil Blackwell, 1972.

# The Anomalous Space of Pictures: Toward a Critique of Stereographic Virtual Reality

Rose Woodcock

> I do not know whether this device will, or can become a medium of art; all that matters in the present context is the undeniable evidence that images can be approximated to the experience of reality.
> (Gombrich 2000)

The design, and accompanying rhetoric, of stereographic '3D' virtual reality (VR) both draw significantly upon natural realism as a self-evident mode for imaging 'reality'. This rhetoric is taken up – uncritically – in a New York Times article on technology, in which a short film *Star Trek: Borg Invasion 4D* is reviewed. This article, like many others in the popular literature and mass media domains, and in films such as *Lawnmower Man* and *Matrix*, positively embrace the idea that computer-generated virtual environments will take the spectator into wholly new and separate realms of experience; if only for short periods of time. Opened at the Las Vegas Hilton in April, 2004, *Borg Invasion 4D* surrounds the viewer with images and sound that create a '3-D movie that not only gives the illusion of a world in front of you, but all around' (Taub 2004). Here, as with many recently developed 3D computer games, users are less spectators of imagery and more like participants inside the action. Oliver Grau, in theorizing virtual reality discourse and practice, uses the term 'image space' (Grau 2003: 13) to describe the nature of immersive 3D virtual works of this kind. The term *image-space* is apposite here, since it implicates vision with notions of 'world' and 'reality': all three are spatially determined modes of experience, and stereographic VR is nothing if not spatial. The notion of *ecstasis* – standing outside of real-world experience – associated with the creation of plausible 'other' worlds in which to immerse oneself electronically, is a potent one in the discourse of VR. Yet the technology which makes these immersive, other-worldly experiences possible, is something to be engaged with first: a head-mounted display, with stereoscopic monitors worn immediately in front of the eyes, provides the key binocular cue for immersion in a virtual world. Stereographic imagery is usually accompanied by audio and even tactile or 'haptic' feedback to enhance the sense (illusion) of separation from the actual world. Thus the 'immersand' is first fully immersed in technology in a very real way. But insofar as stereographic 3D imaging determines the visuality of VR systems, stereography in turn determines that the predominant mode of perceptual engagement in almost all VR worlds will be *visual*. Moreover, VR is specifically designed to accommodate binocular vision and thereby emulate, in the form of stereographic imagery, the three-dimensional structure of real-world phenomena. How objects appear – whether real or virtual – is determined by information-bearing light that enters the eyes. The real world is the default model for building alternative worlds, even where the latter depart radically from it in terms of content, narrative possibility, signification, and context. As a consequence, the real world's *perceptually given* structure – essentially a three-dimensional array of surfaces that structure light entering the eyes – is the measure of the apparent reality of many of its virtual counterparts.

The relationship between the appearance of external reality (as referent), and its translation into imagery through different conventions of representation, is a central theme here. Clearly, not all VR environments emulate the real world as it appears naturally, and many (notably artists more than developers of stereographic technology itself) have challenged the assumption that the virtual should be a response or reaction to it. This chapter is partly a critique of the design and rhetoric of VR as constitutive of a more general, and therefore persuasive and prescriptive, VR discourse: one which reflects and in many ways remains caught up this assumption. But this chapter is largely about vision itself: specifically, vision as a mode of conscious experience. Different conventions of representation excite differing dimensions of awareness from vision. My focus is on how vision is engaged as a means of 'sharing space' with represented objects and entities. This is examined by comparing the stereographic imagery of VR with the pictorial imagery in paintings, respectively.

Grau nominates 'two main poles of meaning of the image' which he sets out as its 'representative function' and its 'constitution of presence' (Grau 2003: 14). For Grau, 'presence' is defined as the 'quality of apparently being present in the images' arising through a 'maximization of realism' associated with virtual imaging technology, and 'increased still further through illusionism and an immersive effect' (Grau 2003: 14). Gombrich's remark is pertinent here. Stereographic imaging reworks the 'approximation' of images 'to the experience of reality', but to the limit of analogous relations: for as the constitutive imagery becomes inseparable from the constituted reality – as the former becomes the complete substitute for the latter – scope for analogy starts to break down.

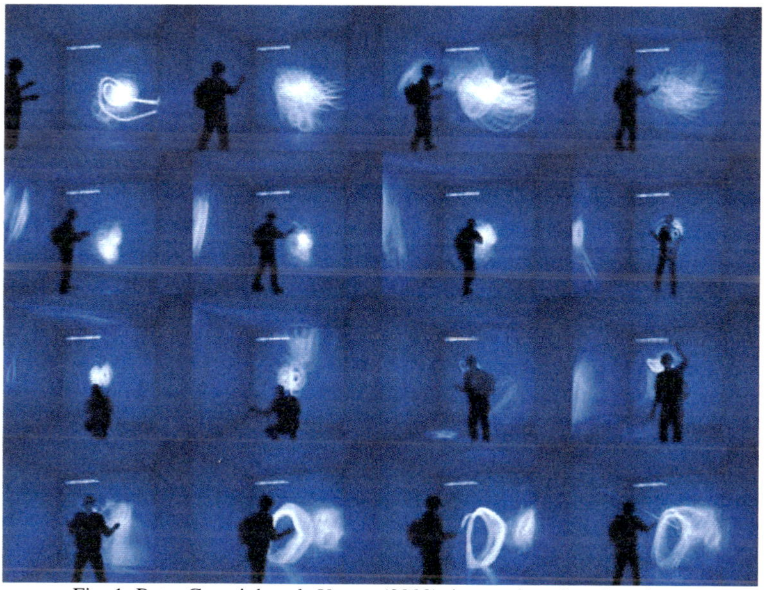

Fig. 1: Petra Gemeinboeck *Uzume* (2002), interactive virtual environment
Image courtesy of the artists

In VR worlds, realism is articulated across two axes: resolution of the imagery, and degree of presence. But not all VR advocates are necessarily uncritical in their application of high-end imaging systems to conceptions of 'being' and 'presence' in visualized spaces. Petra Gemeinboeck designs more subtle, open-ended immersive VR works such as *Uzume* (Fig. 1). She voices a concern about this 'technologically enforced negotiation between the virtual and the real' (Gemeinboeck 2004: 52). Gemeinboeck's is typical of artists whose work challenges and ultimately shapes socio-technical discourses, through reworking the technology itself. But while not all VR practitioners assume a direct correspondence between real world appearance and the inherent ability of this technology to imitate nature, designers and spectators of VR 'worlds' typically remain enamoured of VR's mimetic capabilities. Of the various aspects that can be mimetically given, visual appearance has tacit connections with notions about how reality can be known, experienced, and represented.

New visualization technologies seem poised to bring into being previously unrepresentable worlds, objects and spaces. Yet the convergence of the already powerful spatial imaging system of stereography with that of photorealism, however, marks a return to representational realism, which the plastic and performing arts (to varying degrees) have long, and often vigorously, rejected. Both painting (particularly since the mid-19th century) and photography (since almost its inception[1]) have systematically challenged – through abstraction and expressionism – any presupposed equivalence between artistic purpose or value and verisimilitude. The visual-perceptual possibilities associated with the departure from realism is an underlying theme here, especially concerning the visual constitution of space.

The study of human vision provides a point of entry for attempts to understand consciousness. A comparison between the types of seeing elicited by different modes of visual representation, draws attention to overlooked aspects of visual awareness. The psychologist and founder of computer visualization, James J. Gibson, noted the paradoxical nature of pictorial perception, stating that the 'kind of vision we get from pictures is harder to understand than the kind we get from

---

[1] The earliest known fixed photograph, by Niepce, was in 1826. By 1857 Oscar Rejlander's 'Two Ways of Life' incorporated numerous negative exposures composited to form a complex, multi-layered moral statement about life's choices. Laszlo Moholy-Nagy's 'photograms' similarly included expressive, formalist uses of the photographic medium. Florence Henri, in 1930, began to exploit the anomalous perspectives of mirrors in photographing objects: far from the 'truthful' representation of the real world, Henri's photographic images distorted space.

ambient light, not easier' (Gibson 1978: 227). A picture's capacity to 'bring into being' depicted objects, figures, spaces, and indeed entire 'worlds' – whose ontology is dependent upon pictorial conventions – raises questions not only about the capacity of pictorial representation. Pictures also prompt enquiry into how the human visual system confronts pictures *as* pictures, and not (say) as flat, rectilinear, parti-coloured objects. The focus of this chapter is a comparison between varieties of visual experience. The key distinction is made between the notional space of pictures, and the 'solid' stereoscopic space of VR.

Etymologically, stereo-scopic means 'solid sight'. VR's capacity to assemble solid reality from within vision's own foundations is a curiously new response to the Platonic rejection of the deceptive realm of appearances: for in an immersive virtual world, one not only sees solid forms, but also inhabits that imaginary world as if one were within it. Objects in VR appear fully present to the visual senses even while the spectator knows full well that they are not really there: thus, the ability of new imaging technology to afford a sense of 'presence' for the spectator is associated with an increase in imaginary and sensory possibilities. Yet these possibilities remain grounded in the real-world conditions of binocular vision. The three-dimensional space of VR provides an intense experience of three-dimensional spatial awareness; something pictures cannot do. However, VR simultaneously closes off other perceptual dimensions – such as anomalous or ambiguous spatial cues – of the kind associated with pictorial perception. I propose that pictures afford a mode of spatial awareness that challenges the robust, three-dimensional space of stereographic VR. Pictorial space does not, of course, abide by or necessarily always seek to imitate real-world structure: pictures, as Gombrich (and Gibson) remind us, are 'infinitely ambiguous', even when they do appear to represent the real world (Gombrich 1963: 157).

Does stereographic virtual imagery offer infinite ambiguity too? Or is perceptual ambiguity a quality peculiar to pictures and pictorial conventions, or, alternately, to those instances in real life where we mis-perceive?[2] If, as I argue, stereographic image spaces are not

---

[2] Human vision mostly provides veridical information about real world structure, although it is not infallible and can be fooled. Spatial vision is itself inherently ambiguous: see for example Stephen E. Palmer's discussion of the 'logical ambiguity

pictorial but *real* (or at least *real-like*) even where their content may be highly fictive, then their spaces can hardly be 'virtual'. David Summers defines real spaces as those 'we find ourselves sharing with other people and things' whereas virtual space is 'space represented on a surface, space we "seem to see"'(Summers 2003: 43). What are the equivalent qualities or dimensions of ambiguity and anomaly in VR, if space and the people and things immersed in it are perceptually 'real'?

On a comparison between spatialities, pictures cast the space of stereographic VR as inherently non-ambiguous, or 'solid'. Yet VR does have capabilities beyond its designers' quest toward the realization of real space. Stereographic imaging technology can be reconfigured in ways which push it from the mimetic to something approaching the ambiguity of pictured spaces. The purpose of such a shift would not be merely to reproduce pictoriality in three-dimensions (walk-through Mondrians), nor to place pictures inside 'virtual galleries'. While this chapter sets out to critique discourses and practices of VR which would hold it to 'what it does best', I propose an alternative approach to the design of stereoscopic VR: to have it 'do badly'. That is, to present spatial cues analogous to those in pictures – the discontinuous, irrational, anomalous spaces of Cubism, and abstraction more generally – while simultaneously maintaining the three-dimensional spatial awareness of stereography. These anomalous modes of visual reasoning are already well articulated within the limited means of planar pictorial space discussed below. To introduce them into the immersive stereographic space of VR presents a radical reworking of figure-ground relations, conventions of visualizing spaces, and the relations spectators have with such spaces.

Twenty years ago I spent a long time looking at a painting by Georg Baselitz in the Australian National Gallery, Canberra. *Meissener Waldarbeiter ('Meissen Woodmen')* is a relatively large painting, around two metres square (Figure 2). What was fascinating about this painting was the way it represented space across and within the picture surface. At the time, I was struck by the chaotic relations of figure and ground, and by the impression that the entire contents of

---

of depth information' due to the inversion of 3D imagery from the real world to 2D retinal imagery and back into 3D visual awareness. Palmer 1999: 201–202.

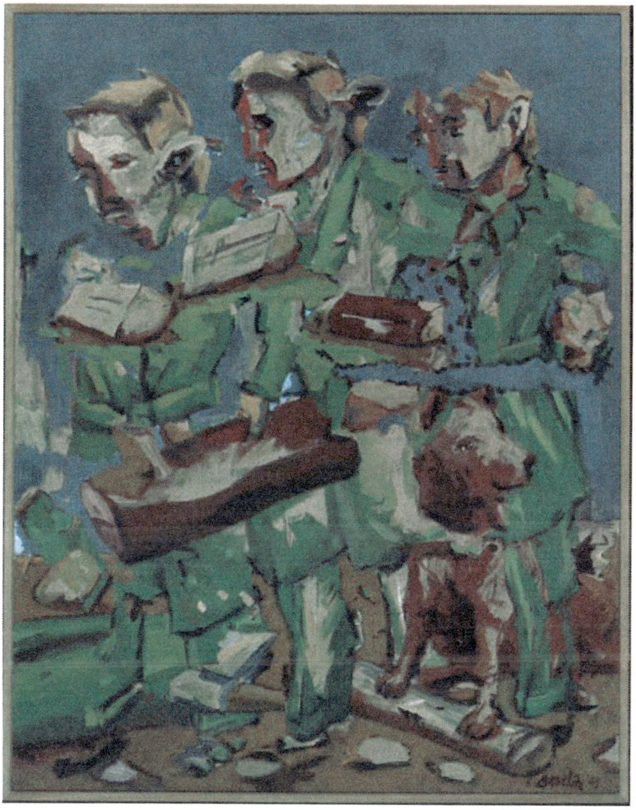

Fig. 2: Georg BASELITZ
*Meissener Waldarbeiter [Meissen woodmen]* 1969
Oil on canvas, 250.0 x 198.0 cm
National Gallery of Australia, Canberra.


the painting – three woodcutters, some cut timber, an axe, a dog, ground and sky – were tumbling forward out of the frame. On a recent visit to the ANG, I was again struck by this painting in much the same way. That *Meissen Woodmen* still affected me like this after twenty years is significant, not because of the longevity of its effect, but because in recent years my interest has been directed to the very differently constituted space of stereoscopic '3D' virtual reality. In

VR, figure and ground are constituted not so much as *representations*, but through actual three-dimensional spatial relations. It is significant that a 'mere' painting, with its limited range of colours, and simple planar configuration, could still elicit a sensation of space that was exciting, ambiguous, dynamic, and extremely engaging. The perceptual play afforded by the anomalous space of this painting is a peculiar and particular dimension of the visual possibilities of pictures.

Pictures, such as Bailey's *Manfroni Still Life* impart to the act of perceiving, a space of ambiguity articulated through inconsistent representation of space: one ought to be able to see down into open elipses. Bailey has manipulated the spatial logic through mixed perceptual cues, each of which are in themselves consistent, but which 'fold' pictorial space along the straight tops of the vessels. Looking at Giotto's *Vision of the Chariot* we see the same curious visual logic, with some sections of the picture plane tipped so as to complicate the entire structure of the scene. Gary M. Radke, writing on the contribution of Giotto to the early developments of expressive painting, notes that Giotto was 'hardly doctrinaire in his perspectival schemes' (Radke 2003: 92). This is not to overlook the constraints of the time of Giotto's painting when, in the mid-1300s, knowledge of perspective rendering was only just becoming a coherent craft, and often only applied to sections of a painting, object by object (across parts of buildings, etc.). What makes these paintings visually so engaging is the fully visualized evidence of the struggle in manufacturing three-dimensional space across the two-dimensional picture plane. The viewing plane – the virtual surface we 'seem to see' in the picture – in this painting[3] is porous, nebulous, manifold, irrational; yet maintains its hold on the contents it depicts. Space in and across the picture plane is mobile: not yet fixed properly by the regime of Cartesian perspective. Moreover, there is a sense in which Giotto contrived some of the impossible, Escher-like architectural structures depicted in various Biblical scenes. Whatever his intentions,

[3] Other paintings by Giotto, such as the *Expulsion of the Demons from Arezzo* and *St Francis Giving his Mantle to a Poor Knight*, have this visual-spatial characteristic. In *Expulsion of the Demons* Giotto makes use of synthetic rather than strictly observational shadows to give localized definition on, for example, certain parts of windows and doorways. The curved, crenulated wall to the right at the very front of the picture is itself composed of mixed perspectival cues.

Giotto's paintings provide a space of perceptual play which, like *Meissen Woodmen* and *Manfroni Still Life*, demands that we attend to how, as well as what, we are seeing.

Virtual imaging technologies and painting share a history of the representation of space. Photorealistic VR arguably has borrowed from the observational sensitivities of painters to the behaviour of light, colour, shadow and form within the framed visual field. The history of painting – both representational and non-representational – is largely an argument about space; the push and pull, into and out from the picture plane through the modelling of form creates the illusion of space as depth in paintings by Vermeer or Caravaggio. The composition of formal elements across planar space creates the infinitely shallow space of Mondrian or Stella. Between them, are the more explicitly ambiguous spaces of Picasso and Cezanne. The art historian Sheldon Nodelman refers to the progressive 'thinning-out of that texture of cues' (Nodelman et al. 2002: 72) by which painters have, since the mid-nineteenth century, challenged the visual conventions of painterly space. From Cimabue to Stella this progression is evident. Yet virtual imaging technologies, as they are typically used, continue to progressively (or conservatively) put 'solid' stereoscopic space back into the image.

The space *as depth* of VR is therefore not illusionistic, and as I have suggested, neither is it virtual. Consequently, the visual representation of space has in a very real way been 'solved' by the advent of digital technologies such as stereoscopic VR, which give the spectator an impression of being immersed 'in' the space of the image as though it were real, not illusory. Between the depiction of space across a picture surface, and the constitution of three-dimensional space through stereopsis,[4] a complete transformation of spatial imagery occurs. This transformation implies a continuation of this progression, from the limited means of pictorial planar representation, to the consummation of space through real (stereographic) three-dimensionality. Except that the progression is reversed. Thus, the conquest of visual space by stereography does not conclude, nor

[4] Three-dimensional vision arising from the lateral displacement of a pair of retinal images from a single binocular view of the visual field. These two slightly offset images are interpreted by the visual system as three-dimensional vision of the scene.

represent the apex of, investigations into the spatial dimensions of visual imagery that began with Cimabue and Giotto.

Thus two modes of spatial interaction can be distinguished: in those cases where *chora* should be translated 'place/space' the idea is always that of an extension, whether two- or three-dimensional, which is occupied or which can be occupied (Algra 1995: 33). Space in pictures can be conceptualized as unused perceptual capacity (Feldman 1980: 16).

Stereographic imaging has distracted us from the spatial possibilities of pictures by positing realism, once again, as the standard by which to measure spatiality (and linked to this, perhaps even 'representational satisfaction'). The rhetoric of VR assumes, and over-determines, a link between 'spatiality' and 'reality' which manifests across two important axes: (1) the design of stereoscopic devices, from hand-held stereo-viewers to sophisticated, immersive VR environments, and, (2) what spectators and critics expect to experience from them. Consequently, designers and spectators of virtual reality systems continue to presume realism to be isometrically related to the three-dimensional structure of the physical world. This presupposition underwrites the thinking and design of new technologies of representation. But what is new about these photorealist, immersive image-spaces, other than that one can occupy their imagery?

In the advent of such technologies, pictures warrant more, rather than less, investigation because of the kind of experiences they elicit for the spectator. One's responses to the spatiality of the Baselitz are structured by conventions peculiar to painting (and drawing); but in an important way are articulated as well by how they depart from the experience of space in the real world: the dialectic of 'pictorial' against 'actual' spaces informs the sense in which we see space 'in' the painting which is nothing like the space of real world perception. The disassembling of compositional space across the picture plane in this painting articulates the image as a whole: the simultaneous confirming and rupturing of the surface, space, and figure / ground relations provokes visual perceptual responses from the spectator. These responses, however, are not an attempt to resolve the spatially illogical visual information for this is a picture, a representation. As such, it does not presuppose to afford real-world space fit for occupancy.

Visual awareness is regarded as a relatively easy aspect of consciousness to study by researchers of consciousness. Visual awareness is ubiquitous, highly structured, it has properties that are easy to articulate, it is relatively easy to monitor, and we have a reasonably good understanding of its basis in the brain (cf. Chalmers). Vision science has shown that certain neural clusters are associated with specific visual events. Whether there are clusters of neurons that deal explicitly with pictorial perception (in a way that would be meaningful to aestheticians) is an interesting, but not central, question. As an approach to understanding vision, visual-neuroscience answers many of the 'big picture' questions concerning human visual awareness: the relationship between external reality (which we presume to exist in some way outside of ourselves), its manner of appearing to us, and how various perceptual and cognitive apparatuses construct such appearances as internalized, real-world knowledge. Computer visualization and computer (machine) vision are sub-disciplines within vision science that continue to provide insight into human visual experience, offering great scope for understanding vision through experimentation and modelling that could not otherwise be done.

It is no coincidence that vision science, especially the study of spatial perception, has been at the core of research into stereographic, virtual imaging technologies. As artifacts that are neither strictly real nor pictorial, virtual imagery occupies a position at the intersection of art and science. New technologies for constituting virtual objects and spaces as perceptual 'actualities' – what I have called 'solid' stereoscopic imagery – recast conceptions about visual awareness and its capacity to apprehend reality 'correctly'. The theme of reality, and its variable representations, is an enduring dialectic across time and cultures. The singular, ontological Reality (such as it exists) is posited against the metaphysical realities of dream, of its narrative forms in art, literature and film, of the realm of the unconscious, the spiritual, and the imaginary. As an object of philosophical and psychological enquiries, reality, and its manifold articulations through art and science, resonates deeply with thoughts about (human) consciousness. Arguably, reality and consciousness are inseparable realities from which we normally cannot – through *ecstasis* – remove ourselves.

Virtual worlds are designed and conceptualized according to the inherent capabilities of binocular vision to render worlds in '3D'. The eyes' capacity to interpret information from the three-dimensional world structured by light, and render it into a three-dimensional reality, is the basis by which most of us sense we know the world and that we can operate predictably within it: it is no wonder that new visualization technologies so avidly set about building complete 'worlds' or 'environments' and not just 'realistic moving images'. Much of the early research into stereographic imaging was instigated by J. J. Gibson. The major departure of Gibsonian theory from previous thinking about vision was the attempt 'to specify how the world structures light in the ambient optic array such that people [on the move] are able to perceive the environment by sampling that information' (Palmer 1999: 53). Gibson's position against the theory of 'unconscious inferences' as being the only avenue for perception led to his advocating the contentious theory of 'direct perception',[5] the theory that all necessary information is available from and fully specified by the dynamic ambient optic array (AOA). Therefore there is no need for inferential processing – e.g., 'mediating processes or internal representations' (Palmer 1999: 54). Unconscious inferences presuppose drawing beyond what is available to the senses from the external stimuli. The ideas of Gibson and his disciples on ecological optics helped validate a new emphasis on mathematical analysis and modelling of the relationship between environmental structure and image structure. Gibson's theory, despite its limitations, did shift emphasis from a 'black box' conception, i.e., where perception arose from its mechanistic basis in the brain, to an informational basis as determined by the structure of the environment.[6] But Gibson was also intrigued by the duality of pictorial perception whereby we see 'both a scene [depiction] and a surface' (Gibson 1978: 233). One seemingly looks beyond what physically appears on the surface of a picture to

---

[5] Gibson's notion of direct perception fails to account for the recovery of real-world information about the external three-dimensionally structured environment from the two-dimensional retinal images (what is known in vision science as the 'inverse problem').

[6] There is still much debate on how the brain solves the correspondence problem of binocular disparity. Palmer states that despite the 'substantial progress that has been made through such physiological studies of stereoscopic vision, no light has yet been shed on the specific algorithms through which the visual system solves the correspondence problem': Palmer, 202–203.

'something else': something that does not exist as such, but is nevertheless somehow 'there'.

It is clear that in computer-generated stereoscopic virtual reality systems, something quite different from picturing occurs. The kind of seeing associated with virtual environments is arguably closer to the ecological model of visual perception put forward by Gibson, than to what is experienced when viewing pictures. It might be stated that this is of course exactly the point of designing VR systems in such a way: precisely to avoid merely producing three-dimensional 'moving pictures' of depicted objects or worlds. In relating stereography to an ecological mode of referencing vision, I argue that the mode of imaging in stereographic VR shifts vision from an 'artifactual' to an 'ecological' mode of visual experience which entails a loss of the dimension of seeing associated with pictures. Seeing pictures as pictures suggests an additional, supplementary mode of seeing along with ordinary, real world perception. Seeing stereographic imagery as constitutionally the same as that in the ordinary world, suggests a reduction, or redundancy. This supplementary nature of pictorial perception is, then, part of the strange attraction of pictures. Pictures do not fit our immediate sense of what the ecological or biological-determinist bases of visual development and function might be.

The dimensions of awareness associated with pictorial representation have attracted contending views – from among vision scientists of differing approaches, psychologists, and aestheticians within different branches of philosophy[7] – about what it is we actually experience, desire, or 'do' when looking at depicted entities in pictures as opposed to real world scenes. Which aspects of visual awareness enable a viewer to 'see' beings occupying space in imagery, where none actually exists?

Human binocular vision's acuity may not be as good as that of other animals, but the human brain, unlike that of other well-visioned beings, can and does manipulate visual experience according to intentions that are not tied to 'ecological' concerns (of pragmatic survival). Pictures can be abstract, consist only of monochrome lines,

[7] For example, Wittgenstein, Scruton, and Wollheim (among others) of analytic philosophy; continental philosophy, notably the phenomenology of Husserl, Ingarden, Sokolowski, Merleau-Ponty, Dufrenne; semiological and post-structural accounts in the writing of Barthes, Derrida, Eco, and others.

consist of a portion of a would-be or entirely fictional world, and yet are considered whole within their frames. We do not normally think of looking for those portions of the depicted world that do not appear within the frame, or that might be hidden behind objects in the scene. However, we do think about their existence in ways that have much to do with consciousness, and imagination. Many of these remarks about what pictures consist of are commonsense. Yet they do raise questions that warrant focused thinking about what we mean by 'reality', or that something is 'there' in the painting. If we think of it as there, then is it not in some sense 'real' to us?

Understanding and theorizing one's relationship to, and the 'reality' of, objects and persons depicted in static images such as paintings, has elicited a range of often conflicting responses from different disciplines. Some thinkers more than others have attributed to pictorial perception a degree of autonomy that sets pictures – or more specifically a peculiar dimension of them – aside from ordinary objects. The phenomenologist Robert Sokolowski argues that there are different kinds of intending such that we use 'perceptual intentions' for ordinary objects, but must 'intend pictorially' with regard to pictorial artifacts. Sokolowski adds that when intending pictorially 'we also have it as a perceptual thing' with pictorial consciousness 'layered upon the perceptual' (Sokolowski 1985: 13). According to phenomenology, in order to recognize pictures as pictures, we need to intend to see them, and their depicted contents, within a manifold awareness of what they are. Intentionality is thus a core term in phenomenology applied to 'the theory of knowledge' of individuals (as opposed to its practical application). Phenomenological intentionality is basically consciousness of something, or the experience of something: suggesting we cannot merely 'be conscious' or merely 'be experiencing' without an object toward which to direct these modes of awareness. From the perspective of phenomenology, our awareness is always directed toward objects as part of the conscious awareness we bring to the objects and entities with which we share the world. As such, pictures offer ideal circumstances for exploring consciousness. Pictures confront perceptual awareness in particular and peculiar (pictorial) terms that are not associated with ordinary real world experience.

Phenomenology[8] has in various ways found it necessary to account for aspects of the visual experience of pictures that cannot be reduced to textual (semiological) explanations. While Umberto Eco and others have sought to argue that a picture, as a 'text' like any other, operates by means of a system of essentially arbitrary signs, other theorists – notably pictorial semioticians – have argued that while pictures do contain signs, this is not their exclusive mode of signification or representation. Goran Sonesson thus situates pictorial semiotics as going beyond characteristically Barthesean analyses of pictorial (usually photographic) images. In an introductory passage on pictoriality, Sonesson writes:

> There has long existed a discipline focussed on the material character of language, that part by means of which language is given to perception (phonetics) [...] whereas we have hardly begun to consider seriously the material, and therefore perceptual, nature of pictorial meaning, let alone the perceptual organization specifically characterizing the pictorial expression plane. (Sonesson 1995: 67)

Pictorial semiotics sets out a methodology for understanding pictures as sign systems, addressing aspects of pictorial perception that cannot be reduced to a pragmatic relation of signifier to signified. As such, pictorial semiotics offers an alternative theory to image analysis that takes into account what pictures are, and how they are situated according to perceptual awareness within a given socio-cultural lifeworld. However, the focus is on meaning (signification). Phenomenological accounts of pictorial perception, however, provide a set of criteria which include attention to the experience of space which the spectator shares, at some level of awareness, with the objects depicted in the picture. My arguments draw from phenomenology to the extent that this approach to pictures emphatically attends to the experience of the pictorial, rather than to just one's reading of its significations. Moreover, the signification of words as compared to the pictorial intending as understood in phenomenology, reveals a further dimension of pictorial intending which is essentially spatial. Words, Sokolowski argues, can be distinguished from pictures according to the different 'directions' of

[8] Through the respective enquiries of Edmund Husserl, Roman Ingarden, Robert Sokolowski, Mikel Dufrenne, and Maurice Merleau-Ponty particularly.

their intending. Sokolowski compares the way pictures intend themselves by bringing the intended object, as he puts it, 'toward me, into my own proximity'. In textual signification, by contrast

> the 'arrow' of intentionality goes through the word to an absent object. It is outward bound. It goes away from my situation here to something else somewhere else. (Sokolowski 1985: 82)

When viewing the real world, our perceptual faculties effortlessly engage us with a seamless array of visual information structured by the behaviour of light on actual surfaces; such vision is consistent, and largely veridical. Between the micro- and macro-levels of the breakdown of optical information that is consciously sorted by the visual system, all information drawn from real world vision is constitutive of 'world-building': there are no parts of the ordinary visual world about which we are likely to say 'that is not consistent with what I am now seeing'. Pictures, as aesthetic objects, present objects and spaces fundamentally unlike real world imagery. Pictures require that we see the depicted world (their subjective content) as ontologically apart from the real world. As Mikel Dufrenne puts it, there is 'nothing but the world, and yet the aesthetic object [the picture] is pregnant with *a* world of its own' (Dufrenne 1973: 149).

Stereoscopic, '3D' virtual environments constitute whole worlds out of imagery that is neither strictly 'pictorial' nor exactly 'ordinary'. Yet, despite offering the viewer fundamentally different visual experiences of their respective contents, paintings and three-dimensional virtual environments are both seen, at least at the basic perceptual level, with the same set of visual capacities: a binocular system that registers and interprets a range of simultaneous monoscopic and stereoscopic visual cues, including colour, surface facture, contrast, perspectival depth, figure-ground relations, motion, and so on. In fact, of all the visual cues listed above, it is only the binocular effect of stereopsis – produced as a consequence of lateral displacement (parallax) of the two eyes – which is truly stereoscopic. All other information, including perspectival cues for representing depth, can be given as monoscopic (single-eyed) information. Stereopsis, though, is singularly responsible for the perception of spatial depth.[9] The range of depth from binocular vision is also restricted to roughly thirty metres, beyond which the disparity between the two retinal images is too weak to register, and everything literally flattens out. This aspect of human vision implicates pictures again in their curious manner of representing spaces: for even a small reproduction of a rendition of a distant mountain (Cezanne's *Mont Sainte-Victoire*), will elicit the sense that the mountain is 'far away' and not somehow fixed at the surface of the page. Pictures clearly engage perceptual experience according to modest, but seemingly potent, means. While it is no doubt true that, within the western canon of visual representation, a relatively small mountain signifies 'far away' to an educated viewer, there is more going on perceptually than merely recognizing the 'sign' correctly. As phenomenological observation suggests, we do have some sense of the mountain in Cezanne's painting as something present to us, existing albeit pictorially, in the *absence* of the real mountain. We do not seek to enter into the real world of the view of Mount Sainte-Victoire, because, in my view, that is not what pictures or pictorial intending do best.

Viktor Sklovski has said that the aim of art is 'to convey a sense of the object, to make us see it, not to [merely] recognize it' and that in art 'the process of perception is an end in itself and must be lengthened' (quoted in Lovink 2006). We can elaborate Sklovski's point by arguing that the visuality of stereoscopic VR, for all its world-building capabilities, has lost its artifactuality – a concept which I have linked to aesthetic contemplation and appreciation associated with this 'lengthened' mode of perceptual engagement. I also use the term 'artifact' because it emphasizes the constitutive foundations of the visual means in either case: the painterly or graphically pictorial as distinct from the stereoscopic and virtual. My interest in the two distinct kinds of seeing afforded by pictures and virtual reality, is what their fundamental differences tell us about the capacity of human visual perception: that it is capable of more than confirming vision's own natural veridical capability.

---

[9] The reliance on binocular vision for spatial depth perception is demonstrated by the difficulty we have in judging near distances accurately (parking a car, catching a ball, or pegging clothes on a line) with one eye closed.

I suggest, then, that we emphatically relate ordinary real world visual experience with the 'solid sight' of stereographic VR. By collapsing these into one kind of visuality, the limitations of stereographic imagery emerge: that is, we posit pictorial imagery as the standard against which to test the perceptual capacity of stereoscopic imagery (rather than the other way around). The capacity for ambiguity, visual anomaly, perceptual play, as described as inherent characteristics of the paintings by Baselitz, Giotto, and Bailey, are dimensions of visual experience which are lacking from the robust, pragmatic vision of stereographic VR. Any object can become or be seen as an artifact, and be afforded aesthetic value. The focus on the 'artifactuality' of pictorial perception is a means by which to identify the peculiar dimension of pictorial representation that does distinguish pictures from ordinary objects. I have also suggested that spatial ambiguity is an important aspect of certain forms of picturing. But of particular interest here is how these dimensions distinguish pictorial representation from the 'presentative' imagery of stereographic VR.

My concern is that presenting imagery in terms of a three-dimensional immersive 'world' via stereography, amounts to rendering the technology itself invisible. Moreover, stereographic imaging obscures from the spectator their own sense of what it is to see absent things and bring them into being, into presentness. In virtual worlds – well constructed ones at least – the viewer can forget about the visual-perceptual apparatus of seeing. But this entails a loss of the artifactual dimension of perception, since it is through seeing pictures as pictorial and not ordinary (actual) entities that enables aesthetic contemplation of the limited means of pictures to bring into being their contents: awareness of the artifactual nature of the object is part of the pleasure in seeing it as something other than what it ostensibly is. R. L. Rutsky expresses the same reticence toward the power of virtual imaging technologies, stating that:

> The very idea of virtual reality implies the transparency of the technologies that produce it. The function of virtual-reality technologies is to allow users to be 'immersed' in the 'reality' that they present, to make that 'reality' as fully present as possible. In order to achieve this experience of 'immersion', virtual-reality technologies must efface their own technological form, make that form transparent or invisible to their users. (Rutsky 2002: 111)

Rendering seamless worlds of which all the constitutive imagery is part of a perceptually homogenous whole is analogous to 'hiding the strings' (as in puppetry). Oliver Grau uses the term 'image space' as a way of identifying, as well as defining, what characterizes the shift from images contemplated as entities dissociated from real-world visual awareness (as pictures), to images entered as spatial 'worlds'. Consequences arise from placing the viewer inside the 'image space'. When the signs (metaphorical 'strings') which would otherwise proclaim that the three-dimensional imagery is a technologically mediated construct, are transparent, the viewer has no space of their own outside the image from which regard the experience: stereography operates on a level in which the visual sense – particularly that of spatial perception – is effectively lost to awareness; collapsed into the visualization technology. That is, the spectator is made unaware of the mediatory apparatus – in this case her or his own binocular visual system – out of which the vivid experience of actual space occurs. By contrast, *Meissen Woodmen* presents a space to be seen, looked at, contemplated for what it is as pictorially given (intended). Baselitz's is a notional, putative space organized around principles of pictorial convention, which, like Bailey's *Manfroni Still Life*, it simultaneously breaks. Here, space is a possibility that derives (returning to Burke's idea) from the 'unused capacity' which is a dimension of the kind of perceiving we do when looking at pictures. It may well be available to real life seeing, as well as the perception of stereographic 3D imagery: yet I suspect it is simply not needed in the rational economy of ecological visual awareness.

The space of the virtual world is one in which the participant is moving, navigating, doing things 'in' the space – the participant (who cannot now be described as a spectator) can do this because the space is perceptually 'there'; or rather the object(s) that occupy that space with the immersand are 'there' in a manner completely different from the way we say the woodsmen, dog, and logs are ambiguously 'there' in the Baselitz painting. Yet the space of stereographic VR, like the space in the painting, is not illusionistic: as argued above, stereoscopic imagery gives us perceptual 'solids'. There is no space (no need) for illusion. Thus, stereoscopic virtual reality unambiguously structures the space of its world in terms of an array of surfaces which structure light, and, therefore, one's vision of the space.

Painting is referenced as an example of the peculiar spatial-perceptual 'play' (possibility) which is absent from the non-ambiguous space of stereographic VR. Stereography's lack of spatial ambiguity – which is its spatial achievement – is also how it delivers perceptual loss rather than gain. That is, the loss of an artifactuality peculiar to pictures and exemplified (though not exclusively) by paintings.

In stereographic virtual 'worlds' the viewer interacts with the imagery. Such 'image spaces' can be immersive or semi-immersive, but are characterized more by doing and being in the space (space as chora) than by looking at or seeing. Even in more conceptually interesting stereographic VR worlds, the emphasis is on the three-dimensionality of the space: an effect that is possible only with stereographic '3D' imagery. As one example (and there are many more by artists who are thinking through these technologies differently) the imagery of *Uzume* is abstract, non-referential and can take the user into realms of (perceptual) experience that are unique to the immersive and interactive capabilities of stereographic VR. This is something that VR does well.

In stereographic VR, visual perception is itself requisitioned to work as an apparatus or scaffold: the imagery effectively structures the image space into something that is experienced as a Euclidean three-dimensional space for interaction, navigation, and 'doing things'. This, I argue, determines the kind of seeing – a limit to the amount of 'perceptual capacity' – available to the participant (as spectator). With the introduction of photo-realism as a resolution standard in VR, visual experience is further limited: on one hand, what is visualized is a complete, seamless reality with no 'space' for negotiating the spatial terms, since these are presented as a physiologically robust three-dimensional percept. On the other, the appearance of objects conforms to the 'reality/realism' paradigm which painting (through expressionism and abstraction) has established as a relative, optional mode of representation, and not an absolute driven by a teleology that would end in hyperrealism. The push toward more and more reality has increasingly emphasized photorealism as an answer to stereographic VR's previous limitations – a push that in turn drives industry and market expectations for this technology, and as a consequence, shapes the discourse of VR.

Increasingly, photorealistic stereographic VR has become the prescribed path toward achieving the holy grail of much (though not all) research and development in computer visualization: a virtual world that is visually indistinguishable from the real one. Even consumer level stereoscopic displays (toys)[10] are typically marketed according to how they 'expand the imagination' or make you feel you are 'really there': the implication being that there is more 'reality' and scope for imaginative pleasure when things are in 3D rather than across a flat, static, two-dimensional plane. Yet Hegel (apparently) warned us that we can only imagine things that are not there (referred to by Gilbert-Rolfe in Colpitt ed. 2002: 125). This observation raises a key point in the critique: that the space in paintings is imaginary in so far as what is present to the visual senses is in reality a flat, parti-coloured surface, yet we see things represented 'there' in the picture. Dialectical relations, between figure and ground, flat abstraction and volumetric naturalism, between the viewer, the physical object that is 'the painting', and the objects and spaces that are depicted but nevertheless 'not there' are foundational to pictorial representation.

Dufrenne's distinction (above) between the world and the world of the aesthetic object suggests that paintings offer a duality of world + world (because we are always ourselves grounded in the world in which we share space with the painting and its contents). By contrast, stereoscopic VR replaces one world with another: there can be no duality here, no dialectic at the level of perceptual experience. A dual-edged trope within VR discourse thereby emerges: (1) There is a presupposition that reality is something that can be indexed as both qualitative (degree of resolution of imagery) and quantitative (how much it extends spatially); and (2) there is an accompanying assumption that this reality can, and should, be known, emulated, and reproduced to the best of existing technical ability.

Yet the achievements of stereographic VR and computer visualization are not trivial: they do offer possibilities for both computer vision and the broader discourse on visuality. However, the key achievement of stereographic VR – the fabrication of three-

[10] ViewMaster was the original manufacturer of this kind of stereoscopic, hand held viewer, marketed for the consumer. The science behind most stereoscopic viewers, including the most sophisticated electronic displays, is practically the same as when Wheatstone began to experiment with 3D optics in the early 1800s.

dimensional 'real' space – is technological in the same sense that human binocular vision is a physiological achievement: the ability to present to the visual sense a complete, seamless reality in robust visual perceptual terms.

The spatial and figurative dimensions of pictures work according to a dialectic of presence and absence – and this is a key to their artifactuality. Something absent is presented (in painting): Something present is presented (in stereographic VR). To represent something absent as present is to operate referentially; indexically (referencing, indicating something 'outside the frame') – as is the case with all pictorial representations. To present something already present as present, in stereographic VR, is to operate in terms of redundancy.

Pictures, their persistence as trans-cultural artifacts over millennia, and the perceptual behaviours they elicit, are a strong argument for there being a heightened (or at the very least, a fundamentally different) visual awareness associated with pictorial perception. One might suggest human vision has an appetite for more than is present to the immediate senses; a non-ecological scope for aesthetic and metaphysical experiences which pictures have the peculiar capacity to give. The aim of this critique of stereographic VR is to examine its foundational grounds and limitations (its '3D-ness') and thereby elicit new and different visual possibilities for stereographic imaging technology. There is nothing to be gained through dismissing stereographic VR and its technological possibilities, as though they offer nothing new to the enquiry into visual awareness and perceptual possibility.

Binocular rivalry – in which slightly different images are presented to each eye so as to present two worlds simultaneously – is one approach to consider. This would entail combining monoscopic and stereoscopic percepts within the same visual field. The 'duoscopic' image space which this implies will be an interesting avenue of pursuit in terms of wresting stereographic technology away from its designers' over-determination to conquer space and our engagement with it.

*References*

Algra, Keimpe: 'Concepts of Space in Greek Thought.' *Philosophia Antiqua: A Series of Studies on Ancient Philosophy*. Ed. J. Mansfield, D.T. Runia, J.C.M. Van Winden. Leiden, New York, Köln: E.J. Brill, 1995.

Chalmers, David J. 'Towards a Science of Consciousness.' *3, CogNet Proceedings* MIT <http://cognet.mit.edu/posters/TUCSON3/Chalmers.Vision.html>.

Colpitt, Frances (ed.). *Abstract Art in the Late Twentieth Century*. Cambridge, England: Cambridge University Press, 2002.

Dufrenne, Mikel. *The Phenomenology of Aesthetic Experience*. Transl. Edward S. Casey. Evanston, Illinois: Northwestern University Press, 1973.

Feldman, Edmund Burke. *Varieties of Visual Experience*. Englewood Cliffs, New Jersey, Prentice Hall, New York: Harry N. Abrams, 1980.

Gemeinboeck, Petra. 'Virtual Reality: space of negation.' *Visual Studies* vol. 19/1. Routledge Taylor & Francis Group, 2004.

Gibson, James J. 'The Ecological Approach to the Visual Perception of Pictures.' *Leonardo*. Pergamon Press Ltd., England, 1978, Vol. 11.

Gombrich, E.H. *Meditations on a Hobby Horse: And other essays on the theory of art*. London: Phaidon Press Limited, 1963 (this edition 1994).

—. Preface to the 2000 Edition: *Images and Signs*, Princeton University Press, appearing in: *Leonardo* Online journal: <http://davinci.ntu.ac.uk/gombrich/leonardo/gom2000.htm>.

Grau, Oliver. *Virtual Art: from illusion to immersion*. Transl. Gloria Custance. Cambridge Mass: MIT Press, 2003.

Lovink, Geert. 'Catalogue of the 1st international video festival at the Bauhaus, Dessau.' *Mediamatic* online journal, special vol. 8/1: *Storage Mania* (sighted June 2006).

Nodelman, Sheldon, Marden, Novros, Rothko. 'Painting in the Age of Actuality.' *Abstract Art in the Late Twentieth Century*. Ed. Frances Colpitt. Cambridge, UK: Cambridge University Press, 2002.

Palmer, Stephen E. *Vision Science: from Photons to Phenomenology*. Cambridge, Mass & London, England: MIT Press, 1999.

Radke, Gary M. 'Giotto and Architecture.' *Cambridge Companion to Giotto*. Ed. Anne Derbes and Mark Sandona. New York: Cambridge University Press, 2003.

Rutsky, R. L. *High Techné: Art and Technology from the Machine Aesthetic to the Posthuman.* Electronic Meditations, Volume 2. Minneapolis; London: University of Minnesota Press, 2002.

Sokolowski, Robert. *Introduction to Phenomenology*. Cambridge: Cambridge University Press, 1985.

Sonesson, Goran. 'On Pictoriality: The Impact of the Perceptual Model in the Development of Pictorial Semiotics.' *Advances in Visual Semiotics: The Semiotic Web 1992–93*. Ed. Thomas A. Sebeok, Jean Umiker-Sebeok. Berlin and New York: Mouton de Gruyter, 1995.

Summers, David. *Real Spaces: World Art History and the Rise of Western Modernism*. London and New York: Phaidon Press Limited, 2003.

Taub, Erica A. 'For the Viewer, No Escape Hatch in Digital 3-D Film.' New York Times web article, May 6, 2004, <http://www.nytimes.com/2004/05/06/technology/circuits/06vega.html>.

# Shifting Figure and Ground in some Australian Photography

James McArdle

I was lost. A mate and I, teenagers, were ambitiously building a hut on a spur in the Lerderderg Gorge and had gone for a pre-breakfast stroll. We found our way back that night, and a weird disorientation, or should I say *relocation* had taken place, so that the site I knew was now somehow *facing another way*. Imprinted for me forever upon this place is this upheaval of the sense of place in space.

White Australian popular history and culture repeats tales of our explorers and children vanishing into the uncanny, hostile bush, like the lone figures in Fredrick McCubbin's painting *Lost* (1886), based on the story of Clara Crosbie (Pierce 1999), extends the theme of the lost child as a metaphor for the settlers' anxiety in being separated from 'Home' (Britain) and left stranded, compounded by the recurring conviction that Europeans do not belong here. Adam Noyce in his film *Rabbit-Proof Fence* (2001) represents the converse, stolen, not lost, children and an indigenous perspective on the landscape. Is this white 'race memory' the source of this thrill or terror in the bush, the undeniable impact of landscape on our psyche?

Knowledge of such places comes about. That is, it involves incursion and excursion. Linear but curvilinear, rotating about the known and expanding it. 'Here' begins with our body, our focal point. Like a compass rose our world swells from it. This is the physical fact of our perspective; our co-ordinates are measured not from an infinite horizon, but from within us. Our body is that dark home, from which we watch the world face to face, we are nestled within its eye sockets, caves in the wall of our own vast continent. The continent is our body that stands behind us, but it is also what brings us 'here', everywhere we are. Accordingly, the concept of landscape or environment as a neutral entity is difficult to support and some kind of self-reflection or transference is bound to occur because we are always 'there' too. Embedded in language in such terms as 'the sheltering rock' is an assumption of purpose in elements of the landscape and of the landscape as an entity, but close analysis of them reveals that it is actually *our* purpose and *our* being. The word mountain contains our act of climbing it. Such are the parameters of the figure-ground which is our self in the landscape.

In photography, the premise of the environmental portrait (which I had practised in earlier work) is that the environment and the subject together form the portrait, implying a reciprocal relationship of environment with the human in an extension of the figure/ground problem (of the kind found in the well-known vase/faces illusion). However, my interest here is not to deal with Gestalt theories, but with the question of the figure in the landscape, as described by the French phenomenologist Merleau-Ponty who presents us with a situationist, as opposed to geometric/scientific, mapping of this spatiality: 'As far as spatiality is concerned, [...] one's own body is the third term, always tacitly understood, in the figure-background structure, and every figure stands out against the double horizon of external and bodily space' (Merleau-Ponty 1962).

When I take a walk in the bush, it is not long before the pure experience of the surroundings is replaced with a concern for finding my way through it, finding my place within it. 'Passage' (2001) traces such a walk on the land [Figure 1].

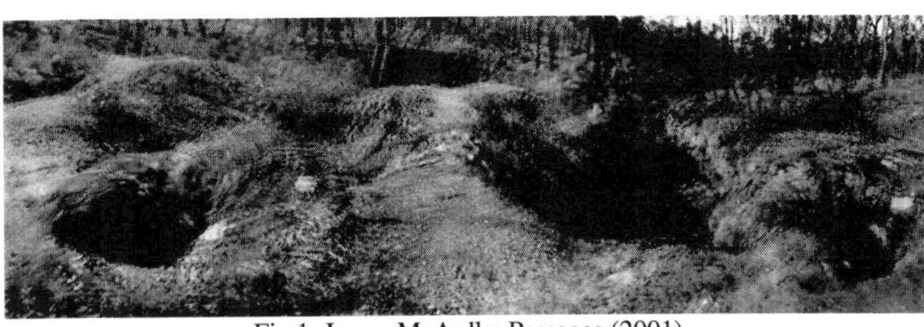

Fig.1: James McArdle: Passages (2001)

It surveys a site that is a pitted, ruined creek bed, turned over by miners several times and sluiced so that the surface is stripped, leaving quartz and sandstone rubble encrusted in a hardened clay that becomes almost fluid in the rare wet months. Treacherous shafts still penetrate the crumbling upheaval of old mullock heaps. The panoramic format and large (5m wide) scale is a strategy that encourages the viewer to 'unscroll' the landscape, and consequently to regard the image as a passage or a journey rather than a 'view'. A viewer who has looked into the pit at either end may realize that they encounter the same shaft at the opposite end of the image, reoriented so it presents two framings of the same hole at right angles to each other. Travelling with the viewer through the panorama and its undulations is a field of moiré patterning that activates the whole surface and spirals into the black maws that penetrate it, then arcing out in amongst the repetition of trunks and grasses. It traces my own experience of this small journey, in venturing out and returning to find my thoughts of other times, future and past.

Traditional perspective is being challenged as we are becoming able to image this other view, where space and location become the same experience. The body, and with it the portrait itself, is *behind* our perception of the landscape (in Merleau-Ponty's sense), its orientation is to perception and the mind as much as the environment.

*Seeing with Both Eyes*

I turn to the problem of properly seeing the 'ground'. For a human being, the landscape is understood in transit. In the process of journeying through the space of the landscape, it becomes a place. The photograph forgets the passage, but it records for us the place, or *presumes* a place, an assumption that it is also a representation of a significant environment – a 'view'. An eye steeped in European landscape looks for vistas and views through bending bough and enframing foliage, but this landscape cannot to be grasped by standing in one place, for how can one snatched moment recreate a sense of place that only the span of time can return?

Might this not be achieved by overlapping and combining these instances? Our two eyes already do this, and thus, as Ernst Mach commented, every person becomes two observers (Mach 1893). By extension comes the possibility that where we stand we are in two places at once and at two points in our movement through space. This is what seeing with both eyes is.

Stereopsis occurs in the brain. The pathway of the output of right and left retinas cross as they enter the brain so that the left hand image is dealt with by the right hand side of the visual cortex and vice versa. Strangely, this crossing is only partial and applies only to the overlapping parts of the field of vision. The peripheral, outer portions of the field remain uncrossed and the right-hand, unduplicated information is processed by the right hemisphere (and reciprocally in the peripheral left field). Cells in the visual cortex thus deal with both duplicated and unduplicated data. Duplicated data is compared simultaneously to highlight the discrepancies caused by spatial separation (Regan 1991).

'Cyclopean' vision is a term used in the literature of human physiology to refer to this effect of seeing one image with two eyes through a 'fusion' of the images. However, it also appears commonly in discussions around mathematical perspective (Panofsky 1991 and Kemp 1997, 1990), often with the connotation that perspective imagery is a construct that signifies panoptic (all-seeing) vision, particularly as arising in Foucault's account in *Discipline and Punish* of Bentham's prison Panopticon. Stereo vision, or stereopsis, coined from the Greek *stereos*, solid or firm, and *oyis*, look or appearance, precedes this fusion into Cyclopean vision.

Binocular vision aroused a great deal of curiosity amongst philosophers from Aristotle and Euclid onwards, who were puzzled that we have two eyes but perceive a single picture of our surroundings. In addition, people with one eye do not perceive a different picture. Early theories of vision held that vision was essentially touch, and naturally the eyes touched the same object, so there was only one impression of a unitary reality (Wade 1998).

Later, experiments showed that an image was projected on the retina (Aguilonius in Wade 1998), and that the sense of depth was provided by the convergence of the optic axes. Using drawings before the invention of photography, Wheatstone demonstrated that the retinal images from two independent sources could be fused into a single image (Wheatstone 1838).

This is a process that is difficult to imagine because it is happening simultaneously with vision, but I believe that the superimpositions that I employ are a graphic representation of it, and that not all forms of spatial representation are calibrations. The stereoscope uses the conventions of binocular vision to produce the effect of three-dimensional vision from two-dimensional images. It re-positions the observer in visual representation, no longer separate from the representation. The representation is situated in fragments outside, but appears within, the body of the observer. Stereo viewing apparatus blends human perception and camera-lens imaging in a unique way that exemplifies my concerns. However, the blend becomes invisible, permitting the illusion that we are looking at three-dimensional space.

Jonathan Crary posits that the stereoscope replaces the camera obscura as the instrument that encapsulates the spirit of its period, citing Descartes's and Diderot's use of the camera obscura as a model for the eye (in Crary 1998). The stereoscope accepted that vision is a function as much of the mind as outside stimuli. This is useful sociologically and philosophically, and prompts a re-evaluation of these instruments for their characteristics in poetic uses.

Photography itself is synthetic perception. The analogy between the human eye and the camera has endured since it was first drawn by C. Scheiner in *Oculus, hoc est fundamentum opticum...* (1619) which he validated in his dissections of animal eyes that permitted him to see the images cast within them, as if they were miniature cameras (as later described by Descartes in *Dioptrique* [1637]) and from which he projected plans for the construction of an artificial eye, which was then built by Rohault (1671). The camera is thus not merely a device, but a construct made with the expectation that it will result in images that are analogous with human vision. The anticipation still exists – Geoffrey Batchen (Batchen: 1997) calls it 'desire' – that the camera obscura, and by implication its evolution into the photographic camera, will replicate and verify what we see.

I recommend we separate the idea of the photograph from the apparatus and connect it with the concept that the process of 'photography' may involve a synthesis of seeing itself, and following from this propose that it is out of the perceptual synthesis that a whole aesthetic branch of the medium grows. My own challenge to traditional perspective is to set up an effect within the image that is a projection of two views, that is, binocular vision. The effect operates in much the same way to disrupt the perspective space by displacing the convention, replacing vanishing points with 'points of apparition'.

Clearly there are limits to the stereoscopic illusion of being 'in the image'. The process also renders voluminous forms as cardboard cutouts, reducing the effect of binocular vision to two-dimensional planes receding in three-dimensional space. What is missing in the information provided to us may be the links our mind makes with other signals from our body about the space we are seeing. Such motion is absent in a stereo image. Part of this feedback is the muscular sensations we receive as our eyes converge on parts of the scene. In fact, our whole biology, our socketed eyes, mobile head, articulated body, participates to satisfy any curiosity about the space in which we find ourselves.

I wanted to see what would happen when the two nearly identical images were simply overlapped, not as an anaglyph (red/green overlapping stereo pair requiring glasses to view). Logically the two views would match up only at one point just as two prints of the same image at different enlargements coincide at only one point. I knew that they would not appear three dimensional, but I guessed that the procedure would uncover something about the way we perceive space. The concept of the new images was to align the images at one place in the scene to imitate the convergence of our eyes which as they focus, converge. A convergence on particulars is evidence of a mind that chooses where to look.

I discovered that the overlap reveals a coincidence in the images representing points equidistant to the observer that are analogous to the convergence of eyes on subjects of attention. The overlapping images also generate a moiré wherever there is sufficient detail on a receding plane. Within the moiré, concentric patterns develop around coincidences between the images so what is revealed is that, what we see, that is, what we attend to, appears at the nodes of a series of vortices. Each vortex is like the whorls of a fingerprint, that is, they are unique in each image, created by the topology of the landscape in which they are created but also by the position of the observer within that landscape [Figure 2], a perspective not with a vanishing point but with a 'point of apparition' that resolves at the nodes of the vortices,

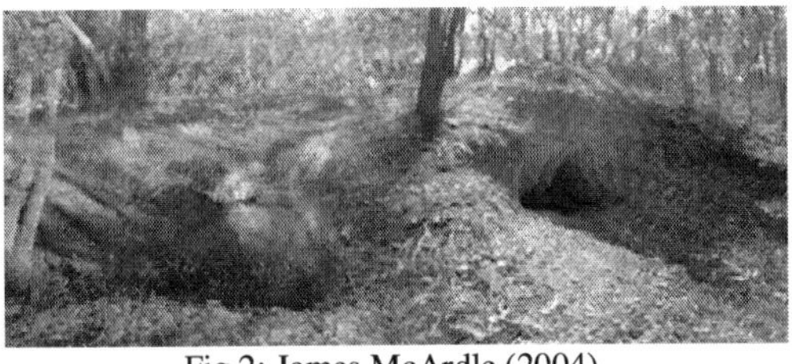

Fig.2: James McArdle (2004).
The exact pivot monochrome inkjet print from large format negatives.

while all other parts of the scene stutter and dissolve in repetition that increases with distance before or behind the point of concentration.

This causes us to reconsider the Renaissance conventions of perspective, which dictate a vanishing point.

For me the vortex reveals a force field in the landscape, the spirit of this landscape, the deep convolutions that have formed it and the history that has tortured it.

*Vision in Motion*

Vision in motion, and binocular and stereo perception are our biological ways of knowing the world. Pictorial, planar imagery is a construction that we can build with mathematical formulae, or by sectioning space with a window, through which we can infer spatial relations. Beyond the binocular construct, a further way to convey spatial information in 2D images is through motion of the observer/camera.

In Jacques Henri Lartigue's famous photograph of an early car race at Le Mans, [Jacques Henri Lartigue (1912) *Grand Prix 1912*] the spectators and car wheels are stretched diagonally in an almost cartoon-like representation of speed and a reaction to it. His clock-work driven camera shutter blinds sliced vertically across a large format negative, consequently images moved relative to the film as they were projected on its surface during a panning shot. He did not quite keep up with the pace of the racing car as it passed stationary spectators, thus car wheel becomes an ellipse that leans to the right while the figures lean to the left. His camera records the motion of time and space.

Daniel Crooks is a Melbourne artist who employs the term 'Time Slice' to describe his images. This was the title of his exhibition at the Centre for Contemporary Photography in Melbourne in June 14 – July 6 2002 (Crooks 2002). He used video to slice the images instead of a static shutter. Each scan is a keyhole in time, a scalpel fashioned from the second hand of an analogue clock, which pares away at the motion in front of it, whether the motion is of figures and vehicles in front of a static camera or whether the camera itself is moving. Crooks's modified camera includes both kinds of motion in one image as it surveys the interior of an elevator, a paranoid claustrophobic micro world in which figures are brief captives. The lift and the camera scan the vertical interior of the building. These time/position 'graphs' are vertical format images with both a time (duration of elevator ride) and position (position of elevator in building and figures in lift enclosure) on the vertical axis and the relation of the travelers in the lift and its doors recorded on the short horizontal axis. Figures entering and leaving the lift as the doors open and close on different levels are recorded as ribbons that thread themselves through the space-time continuum in and out of our field of vision. In *Elevator No.1*, the lozenges of light that punctuate the tall vertical are doors opening and closing, enlarged vertically in proportion to a longer period spent open, while figures become plaits as they weave past each other before settling at either side of the doors.

It is Crooks's choice of subject matter that makes these images far more than an exercise in calibration. The work has precedents in the interest of the Futurists and others in a poetic representation of movement, especially in the urban environment. Crooks achieves a simultaneity that would be the envy of such artists. He positions us in relative time and space inside lifts, trams, trains where local events are nested inside the energies of the city. These are vehicles (in both senses) for a dynamic that is at once strange and familiar.

Inspired by his example my understanding of binocular vision was ramifying into two branches. I understood that with our two eyes we stand at two places at the one time, but now I could see new potential to be derived from the idea that, in movement, with our two eyes we might exist in two moments simultaneously.

*Motion Perspective*

When we first pick up an object, we turn it in our hands so that our sight and sense of touch are exposed to every part of it. The same kind of inspection is extended from this bodily scale into the whole environment as we interact with it. John Herschel was the first to note the effect of rotation in motion perspective: 'Let any one travelling rapidly along a high road fix his eye steadily on any object, but at the same time not entirely withdraw his attention from the general landscape, he will see, or think he sees, the whole landscape thrown into rotation, and moving round that object as a centre' (Herschel 1833).

This is part of a range of phenomena arising from visual kinesthesia. That is, a sense of centric motion that spreads outward from the direction of our travel. J. J. Gibson, who throughout his writings refers to the primacy of motion in perception, calls this the 'optic flow' or 'flow perspective' (Gibson 1979) relating vision to his 'ecological psychology', a theory that recognizes reciprocity between animal and environment.

Depth perception, and a sense of the volume and presence of objects, arises not only from binocular, stereo vision, but also from the active or passive motion of the head, peering around things to get a sense of their proportions and position. Our first experiences of space as a helpless baby are of being carried about, and of course, it is then that the connection between our vision and our environment is established, with the eyes hard-wired to respond to motion, as can be observed in the youngest of babies whose closed eyes 'follow' their surroundings.

Apart from rotation, another word stands out in Herschel's astute observations (Herschel 1833) where, using the word 'attention' he instructs that we need to be aware of the whole landscape before us in order to observe the effect. Herschel emphasizes that the observer has to put themselves into at least two states of awareness in order to be conscious of the rotational effect that arises from an unconscious background awareness of motion. The act of attending to one object, fixing our eyes upon it is not sufficient to notice the effect, though that is an essential part. It requires another level of quite conscious attention to make the surrounding motion stand out even though it is attached to fixation upon an object we are passing.

A connection between motion and perception is at the heart of my research. The effect, called motion perspective, is familiar in the phenomenon of the moon appearing to follow us, steady on the horizon, as we move by in a speeding vehicle. This perception is, as my imagery might reveal, complementary to the vortex pattern created by convergent binocular vision.

My introduction to this effect came in 1995 when I saw Susan Purdy's exhibition 'The Shaking Tree' at Switchback Gallery at the Gippsland campus of Monash University. What I saw there confirmed that there were new ways to work with space in the two-dimensional image. Susan Purdy encountered a continually changing landscape which she photographed as it panned moment-to-moment past the train window. In this case, the motion perspective effect was represented in blurred, relatively slow shutter speed images. The process clearly was an intuitive one, the motion perspective effect at work produced compelling and beautiful images. The result was for me a very powerful evocation of the way we see.

The train is a classic platform for the observation of motion perspective. The passenger is seated, relaxed, at ease. In fact, they are stationary. Through the train window, they observe a world in motion, and yet it does not move any more than the passenger does. Both move in relation to the other and yet the impression of the passenger is that the world outside is somehow frozen, sliced out of time. Philosopher Michel de Certeau (de Certeau 1984) sees a peculiar stillness which attach to his notions of subjectivity. 'A travelling incarceration. Immobile inside the train, seeing immobile things slip by. What is happening? Nothing is moving inside the train or outside the train.' No wonder the carriage provided the most convenient simplification for Einstein in explaining relativity (Johnson 1982). Such observations, repeated by a population of travellers, soon led to an expression of its emotional effects by more astute train passengers, among them poets and artists. Paradoxically, it is the silence of these things put at a distance, behind the windowpane, which, from a great distance, makes our memories speak or draws out of the shadows the dreams of our secrets. Paul Verlaine's poem *La Bonne Chanson II* (1869), evokes not only the way we see in motion perspective from a moving train, but also demonstrates its emotional potential. It is one of the first poems in any language that describes such a scene, a vision which causes the poet so much joy is a projection upon the pivoting point in the landscape on which his eyes rest, the only constant at the centre of 'le tourbillon cruel' which catches up the whole landscape. In this case, it is also a projection of the constancy of his yearning for his beloved Mathilde Mauté, the twenty-six year old poet's wife of barely sixteen, whom he left not long after for his lover Arthur Rimbaud.

This phenomenon of motion is also recounted in the Australian Xavier Herbert's writing, though he uses it, like Purdy does, to express his sense of the alien unknown of the outback landscape: 'seeing the stunted trees, the mulga and the wilga and the gimlet gum, doing a kind of dance, spinning past, seeming to swing away from the train to the horizon and race ahead, to come back to meet us and go waltzing past and round again, the same set of trees in endless gyration, trees that danced a wild arboreal polka to our going' (Herbert 1963).

Where do the internal and external meet? In the train, with a camera [Figure 3], I could be detached from the passing landscape, yet hold in my hand the power to see it all and to track its passing. The other potential is to engage with the motion itself to reflect into it the condition of the traveller in space and time. This I took to the next stage of this research which involved using train travel as a means of producing images that explored another aspect of the portrait in the landscape. The portrait subject in this body of work was the unseen traveller who at the same time is every observer, the portrait image was the landscape which is contains every place.

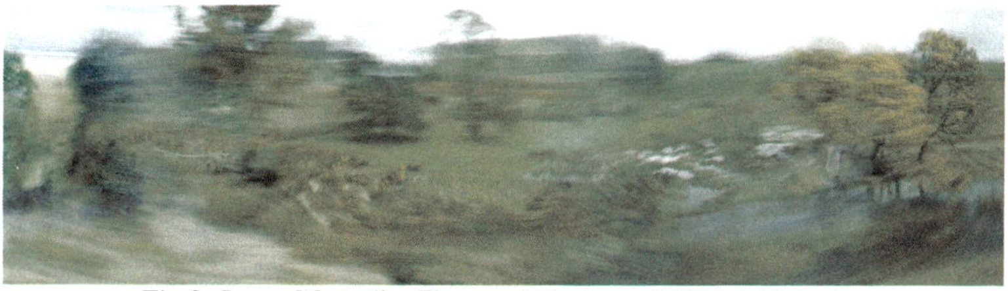

Fig.3: James McArdle: Thousand Horizons (2004) inkjet print

All of these observations gave rise to my question, what does photographing this effect do to our understanding of awareness and attention? I decided to take up the results of Susan Purdy's intuitive photographic technique. Panning on subjects by the roadside produced some results that resembled Purdy's 'Shaking Tree' series. When the subject of the pan, that is the object on which the camera is rotated to keep it 'still' in frame, was a standing form like a tree it could be rendered as an almost static vertical. There was a convincing sense of motion and kinetic forces in these images, a 'spin', just as described by Xavier Herbert (Herbert 1963). What resulted were images in a panoramic format joined with each other to represent the gathering and fleeting of notions, memories and reverie of travelling in this landscape, as we ourselves are gathered up in its motion.

*The Vortex of Vision*

The figure that emerges from this practical research, in both the binocular and the motion perspective imagery is the Vortex itself. As a form it has a long history of association and resonance with things natural, aesthetic and spiritual and equal prominence in mathematical and scientific discovery. There are instances of this structure that have been accorded a visionary status in art and science.

The form of the spiral, whirlpool or vortex, and the related Labyrinth, appear throughout art and literature and are also mystical symbols well known in occult circles. It was an obsession of the eighteenth century, whose thinkers ascribed it ideological, religious, artistic, and moral as well as technical meaning and it is with such traits that it appears in William Blake's Milton (Blake). W.J.T. Mitchell comments on this passage by noting that, 'the Vortex serves as an image of the gateway into a new level of perception', for 'the infinite does not reside in an obscure, transcendent realm at the "vanishing point" of three-dimensional space, but is located immanently in the intense, dialectical perception of immediate "minute particulars", a process which is symbolized and embodied in the vortex' (Mitchell 1978).

Kevin Cope traces the pedigree (Cope 1992) of Blake's vortex from Descartes's writings on natural philosophy as articulated in *The World* (1674). According to Cope, the Cartesian universe is an immense, but knowable, space. Its variegated density is a series of rarefied vortices rotating upon intense knots, encompassing both planetary and atomic scales that exert centrifugal and centripetal forces upon other vortices in an interlocking mechanical system. This system encompasses Descartes's explanation of vision and colour, which he suggests arises from the vortex of an object resonating with the vortex of the eye (*Dioptrics* 1637).

Jonathan Crary (Crary 1999) links Blake and Cézanne with reference to Cézanne's 'sustained attentiveness' when he says: 'William Blake and Cézanne shared a related understanding of the universe as perturbations and differences between centers of energy.' It is here that I find some accord with aspirations for my work in which curving, spiral, helical and vortex forms arise as a reading of the way this compelling form derives from the processes I have used. Let me clarify my position by comparison with David Stephenson's Starlight series.

I first saw them in a major mid-career retrospective 'Sublime Space: David Stephenson Photographs 1989–98' at the National Gallery of Victoria. The Starlight series were metre-square prints. Their content, all arcs and curves and spirals, could seem to be the product of an obsessive geometer, and this impression was reinforced by the edge-to-edge grid-pattern presentation of these prints. The effect of Stephenson's imagery is entirely original, as the viewer comprehends that these images are the recorded passage of stars. The concentric arcs are familiar from astronomy books and camera clubs, but what is seen here is much more complex. The arcs intersect with others, inscribing the blackness with hair'sbreadth curving lines that in the colour prints are an astonishing prismatic hue. Sometimes the arcs are broken into steadily increasing intervals and angles, arrayed to form concentrated hatchings. In '1902' [Figure 4] and others like it, the thatch of short strokes and dots models, in white relief against deep space, a contracting spiral. It is not a galaxy but the abstract for one.

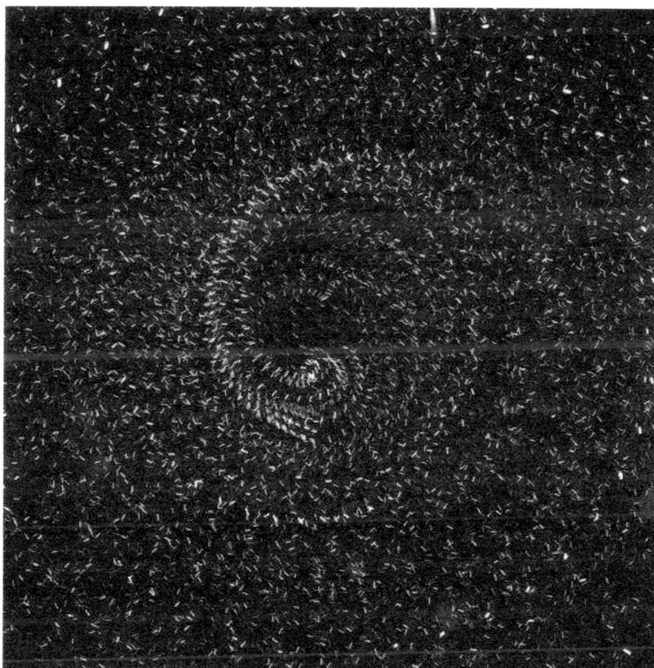

Fig. 4: David Stephenson 1902 (1996)
Chromogenic process colour print
1000mm square from the 'Starlight' series.
Courtesy of the artist.

Stephenson has had to section his nocturnal exposures systematically, in some cases also repeatedly reorienting the camera and tripod to subdivide angles, precise to the minute-arc, relative to the passage of the stars across the sky. The luminous knot of the vortex emerges from the interaction of two time frames, that of the camera and that of the stars and earth. The outcomes transformative not mechanical, the ethereal predominates over the mathematical in these complex geometrical harmonizations, and they are more like mandalas, a meditational orrery with a lineage that can be traced from Descartes's orchestration of the vortices to a transformational modernist torque. In this process, the original star trails and their underlying logic do not entirely vanish but they become abstractions with intimations of the infinite in a re-ordered constellation. These meditational devices are indeed made with traces of the stars themselves, but mediated by Stephenson's calculated re-configuration, so that a design emerges and reorders chaos into a two-dimensional form, a spiritual dimension that resides in a generated harmonic overlay.

The initial element of invention in my investigation was to devise the means by which the process of binocular perception might be depicted. Once the vortex form emerged from that experimentation, and I had the experience to predict the generation of effect, it became possible to manipulate it purposefully in seeking a solution to the problem of the portrait in the landscape.

The observer may be depicted in the photograph as an illustration or document of their presence. They are represented in the 'third person', becoming, when pictured, an 'object' attached to our comprehension of personhood, but then they are not necessarily an observer. For an observer is an entity quite distinct from a viewer, who, in any image, is a being with open eyes, while the observer can only be identified from their state of mind, their attention. How to put an observer into an image while simultaneously depicting our knowledge of them becomes problematic. Though the observer may be inferred as sharing the same vision as a viewer in the image, they cannot themselves appear. The solution to such the figure/ground paradox can be resolved in the figure of the vortex. [Figure 5]

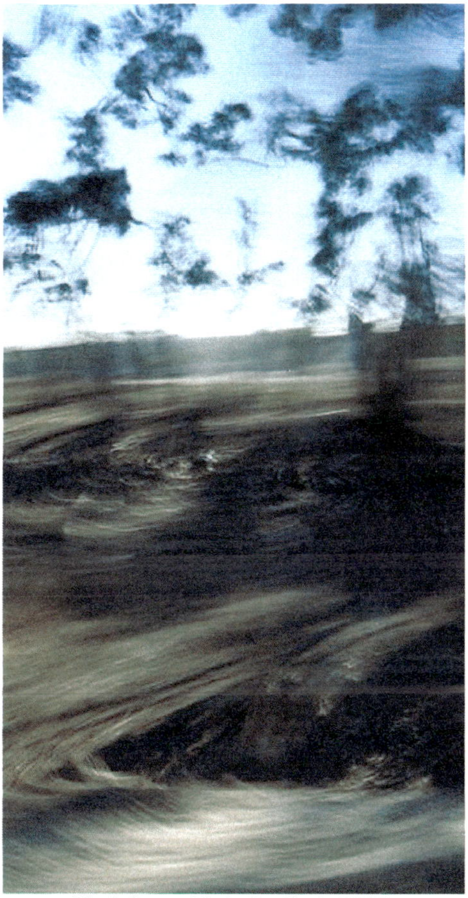

Fig.5: James McArdle: Gush (2007)
Chromogenic print.

*References*

Bachelard, G. *The Poetics of Space*. Transl. E. Gilson. Boston: Beacon Press, 1964.

Certeau, M. de. *The Practice of Everyday Life*. Berkeley: University of California Press, 1984. 111–114.

Cope, K. *Spiral Symmetry*. Vol. 1. Hargittai, I. and Pickover, C. (eds). New Jersey: World Scientific Publishing, 1992. 399–440.

Crary, J. *Picturing Science, Producing Art.* New York: Routledge, 1998. 475–499.

—. *Suspensions of Perception: Attention, Spectacle, and Modern Culture.* Cambridge, Mass.: MIT Press. 1999

Crooks, D. *Timeslice.* Melbourne: Centre for Contemporary Photography, 2002.

Foucault, Michel. *Discipline and Punish: The Birth of the Prison.* London: Penguin Books, 1997.

Gibson, J. J. *The Ecological Approach to Visual Perception.* Hillsdale New Jersey, London: Lawrence Erlbaum Associates Inc., 1979.

Herbert, X. *Disturbing Element.* Melbourne: F. W. Cheshire, 1963.

Herschel, J. F. W. *A Treatise on Astronomy.* London: Longman, Rees, Orme, Brown, Green & Longman, 1833.

Johnson, T. *Insignificance.* London: Methuen in association with The Royal Court Theatre, 1982.

Julesz, B. *Dialogues on Perception.* Bradford: MIT Press, 1994.

Kemp, M. 'Science, Non-science, and Nonsense: The Interpretation of Brunelleschi's Perspective.' *Art History* 1.2 (1978): 134–61.

—. *The Science of Art: Optical Themes in Western Art from Brunelleschi to Seurat.* New Haven: Yale University Press, 1990.

Mach, E. *Popular Scientific Lectures.* vol. 1. Chicago, Illinois: The Open Court Publishing Co., 1893. 66–88.

Merleau-Ponty, M. *Phenomenology of Perception.* London: Routledge and Kegan Paul, 1962.

Mitchell, W. J. T. *Blake's Composite Art: A Study of the Illuminated Poetry.* Princeton, N.J.: Princeton University Press, 1942–(1978)

Panofsky, Erwin. *Perspective as Symbolic Form.* Transl. Christopher S. Wood. New York: Zone Books, 1991.

Pierce, P. *The Country of Lost Children: An Australian Anxiety.* Cambridge: Cambridge University Press, 1999.

Regan, D. (ed.). *Binocular Vision.* London: Macmillan, 1991.

Wade, N. J. *A Natural History of Vision.* Cambridge, Mass.: MIT Press, 1998

Wheatstone, C. *Philosophical Transactions of the Royal Society.* 1838.

# Schoenberg's Hat: Objects in Musical Space

Barry Empson

Geometry makes visible the musical consonances.
Boethius

Schoenberg said in a lecture that the tone-row, that sequence of notes which would present the 'musical idea' he wanted to develop, was the same even when it was presented in different orientations. He demonstrated his point with a hat, manipulating it so that it could be seen from above, below, behind, or in front. In every orientation it is still the same hat.[1] Schoenberg made similar comments in his published essays. 'Just as our mind always recognizes for instance, a knife, a bottle or a watch, regardless of its position, and can reproduce it in the imagination in every possible position, even so a musical creator's mind can operate subconsciously with a row of tones, regardless of their direction, regardless of the way in which a mirror might show the mutual relations, which remain a given quality' (Stein 1975: 223). Schoenberg seems to think of a sequence of musical sounds as some sort of object. These 'musical objects' are held to inhabit 'musical space'. That term is deployed often in Schoenberg's writing about music and 'musical space' is elaborated into a densely complex concept which has provoked a great deal of discussion.[2]

Schoenberg was using the hat to explain one aspect of his own musical practice. As an example, we could consider what he does in the piano concerto Op. 42. The twelve-tone row is stated straight away in four basic orientations. The original row is first heard in the piano

---

[1] Reported by Erwin Stein in 'Neue Formprinzipien', *Musikblätter des Anbruch 6. Sonderheft: Arnold Schönberg zum 50. Geburtstage*, 13. September 1924, 295. Quoted in Busch 1985: 7.

[2] Regina Busch has reviewed a great deal of the literature on Schoenberg's concept of musical space (Busch 1985, 1986a, 1986b). Cf. also Morgan 1980.

alone. It is then played backwards; thirdly it is played simultaneously backwards and upside-down, and finally upside-down. This is rather complex, perhaps only detectable with a careful perusal of the score; a simpler, more easily hearable example is the famous 'mirror point' in *Lulu*, the opera of Schoenberg's pupil Alban Berg. At the turning point of the opera, Berg reverses the music – the same notes in the opposite order:

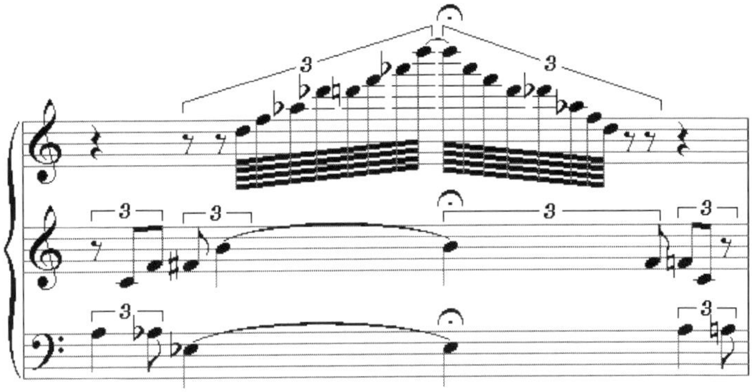

The composer Sir Michael Tippett objected to Schoenberg's hat comparison, arguing: 'A musical tone-row is a succession in time – and the spatial terms, especially that of backwards, are in matters of time fundamentally improper' (Bowen 1995: 31).[3] Tippett understands music in terms of time. He wrote, for example: 'Because music is concerned not with space but with time, this method of artistic creation seems to bypass the problems of representationalism [...]' (Bowen 1995: 10). For Tippett, music 'offers images of the inner world of feelings perceived as flow'. Setting aside the question of what Tippett means by 'images' and for the moment leaving open the possibility that the term 'flow' is already a spatial metaphor, it is still not difficult to feel the force of his objection to Schoenberg's explanation. If music is concerned with time and has nothing to do with space Schoenberg is speaking nonsense.

---

[3] It seems odd that Tippett should single out 'backwards' for comment, when 'upside-down' is even odder.

The great aesthetic debates of the eighteenth century, in whose echo we still converse on these matters,[4] already address a tension between space and time. Lessing argued that painting is an art of space, where poetry (*Dichtkunst*) is of time, and deals with action (*Handlung*). When Herder entered the discussion, he argued that while painting is a spatial art, the art associated with time is not poetry but music. He put it succinctly in the first *Kritisches Wäldchen* of 1769: 'Malerei wirkt ganz durch den Raum, so wie Musik durch die Zeitfolge' (Herder 1813: 183). Herder still counts music among the mimetic arts, however. Its 'objects' are not objects as usually conceived but rather those 'Dinge und Vorfallenheiten, die vorzüglich durch Bewegung und Töne ausgedrückt werden können' (216).

The importance of music in the theorizing of the young German Romantics in Jena is well attested.[5] They continued to express the view that music is of time, and its significance for them arose from its connection with inner sense and from its very transitoriness and insubstantiality, its temporality. Hearing was for Friedrich Schlegel 'the most noble sense' (*der edelste Sinn*) because hearing 'as the sense for what moves (*das Bewegliche*) is more closely connected to freedom and to that extent more suited to freeing us from the hegemony of things than all the others' (quoted in Frank 1989: 298–299). Schopenhauer in the section of *Die Welt als Wille und Vorstellung* on the senses asserts that what the ear perceives is exclusively in time, and that the essence of music consists in tempo (*Zeitmaass*). Hegel also sees music as an art of time. Music 'negates' (*aufhebt*) space and appears as a temporal ideality: 'sound, so to speak, liberates the ideal content from its immersion in matter' (Inwood ed. 1993: 95). Music dispenses with the spatial but retains temporality. We could add similar views from more recent discussion. Stravinsky, for example, claimed in his *Poetics of Music* that 'music is a chronologic art as painting is a spatial art' (Stravinsky 1970:1).

Talk about music was full of spatial expressions long before Schoenberg and despite the insistence on its being a 'chronologic art' remains so. Many of these expressions are familiar from ordinary talk

---

[4] Cf. e.g. Jameson's hypothesis that spatialization undermines our capacity to think history: Jameson 1984. Cf also Fugmann 1998, where Fugmann argues the temporalization of space in the work of Anselm Kiefer.

[5] See for example Bowie 2000: 241ff.

about music. A melody rises and falls, harmonies may be close, as in a barber-shop quartet, or well-spaced. Certain chords in western music are held to be 'unstable' while others are 'stable'. We apprehend in the complex of sounds music offers us a foreground – often a tune – and a background – often harmonic. In the technical language of musical discourse a note is said to be the 'root' or 'base' (in German *Grundton*, 'ground tone') of a chord. Often the language used to describe a melody refers to its 'shape': it may be square, or broad, or an arch, and so on.[6] Marion Guck has suggested that some of the spatial terms used are 'music-literal' (Guck 1997: 201), words like 'high' and 'low' of pitches; others are still felt to be figurative, the application from one domain (in our case space) to another where it does not 'literally' apply.

The spatial terms used in discussion of music suggest differing views about the qualities of the space they imply. As long ago as 1558, the theoretician Zarlino wrote about polyphonic music:

> Just as the earth is the foundation for the other elements, so does the bass have the property of sustaining, establishing and strengthening and gives increase to the other parts. It is thus taken as the basis and foundation of harmony and is called the bass – as if it were the basis and support. But just as would happen if the element of the Earth were missing (if this were possible), what beautiful order of things would be ruined, and earthly and human harmony be destroyed, just so, if the bass were missing, all the *cantilena* would be filled with confusion and dissonance; and everything would collapse in ruin.[7]

However self-conscious this 'as if it were' may be, Zarlino's description of the function of the bass is spatial. The image recurs in Schopenhauer, for all his denial that music is of space. He tells how he

---

[6] Charles Rosen, for example, uses the terms 'broad' and 'square' of a Beethoven melody, although he too is of the opinion that music is not spatial. Rosen 1970: 380. Cf. also: 'Too often the music could be played backward without affecting the analysis in any significant way. This is to treat music as a spatial art' (40).

[7] 'E si come la Terra è posta per il fundamento de / gli altri Elementi; cosi il Basso hà la proprietà, che sostiene, fortifica, & da accrescimento alle altre parti; conciosiache è posto per Basa & fondamento dell'Harmonia; onde è detto Basso, quasi Basa, & sostenimento dell' altre parti. Ma si come averebbe, quando l'Elemento della Terra mancasse (se ciò fusse possibile) che tanto bell' ordine di cose ruinarebbe, & si guastarebbe la mondana, & la humana Harmonia; cosi quando'l Basso mancasse, tutta la cantilena si empirebbe di confusione & di dissonanza; & ogni cosa andarebbe in ruina' (Zarlino 1573: 281–282).

recognizes in the bass in harmonic music 'die niedrigsten Stufen der Objektivation des Willens', and compares these lowest steps to 'die unorganische Natur, die Masse des Planeten'. The 'higher' voices in harmony figure higher aspects of the 'objectivation' of the will until we reach the highest, in the melody:

> Endlich in der Melodie, in der hohen singenden, das Ganze leitenden und mit ungebundener Willkür in ununterbrochenem, bedeutungsvollem Zusammenhange eines Gedankens vom Anfang bis zum Ende fortschreitenden, ein Ganzes darstellenden Hauptstimme, erkenne ich die höchste Stufe der Objektivation des Willens wieder, das besonnene Leben und Streben des Menschen.[8]

Between Zarlino and Schopenhauer, Rameau developed a theory of harmony analogous to a theory of gravity. He was called by some contemporaries 'the Newton of harmony'. In his system there is a note, the key note, which acts as a gravitational centre, attracting all the other available notes. With his investigation of 'corps sonore' and the ratios of resonating partials he thought he had established laws of nature applicable to harmony, because the notes of the triad are constructed from these very same ratios. By 1737 his 'Newtonian' theory of harmony was established.[9] According to several commentators his theories continue to 'form the foundation of most pedagogies of music'. In the western harmonic tradition, a key is established, having a tonic note which acts as a central reference to which all other keys and notes are subordinated: a centre of gravity to which harmonies and melodies are attracted. We could pursue spatial implications of Rameau's work further into the distribution of the notes that sound simultaneously in a chord. These chords can be 'inverted' – Rameau speaks of 'renversements' – when the lowest sounding note is transposed up an octave. The 'fundamental note' is not heard in the bass, but remains implicit. Since he is speaking of triads, a second inversion can occur, leaving the fifth sounding at the bottom. These three are all the same chord, in different orientations – and become increasingly unstable. (Schoenberg's 'inversion' is different: instead of moving the notes of a chord he inverts the in-

---

[8] The quotations from Schopenhauer are all from §52 of the third book of volume 1 of *Die Welt als Wille und Vorstellung*.

[9] 'Génération harmonique', 1737.

tervals between them. Where Rameau would invert C–E–G to E–G–C, Schoenberg would invert the rise of a major third to the descent of a major third, the rise of a minor third to the descent of a major third and by inverting C–E–G would produce a triad in f-minor.)

Hegel, despite all that he says in his *Aesthetics* about music negating space, talks about what he hears in music in terms that are decidedly spatial. A few pages after his statement about 'das Aufheben der räumlichen Objektivität' (Hegel: 260) in music and his discussion of music's 'gegenstandslose Innerlichkeit in betreff auf den Inhalt wie auf die Ausdrucksweise' (262) he waxes lyrical about certain musical effects, using a range of spatial expressions: 'Ausweitung, Verbreitung, ein Auseinandergehen, eine Entfernung und Zurückfindung' of voices (267). He writes of sounds being 'in sich selbst eine Totalität von Unterschieden' which can 'sich entzweien und verbinden' (273). He speaks of the entrance, striding forth, struggle, and disappearance (*Verschwinden*) of voices. In harmony he uses the familiar image of the 'substantielle Basis', and 'der gesetzmäßige Grund und Boden' (298). Melody is said to 'hover above' metre, rhythm and harmony. Of instrumental music he speaks of a purely musical forwards and backwards (*Hin und Her*), up and down (*Auf und Ab*) of harmonic and melodic movement. Lines of music seek and find each other (326).

Tippett himself is said to construct music in 'blocks'. Mark Morris describes the structure of Tippett's third symphony: 'a two-movement structure combining aspects of sonata form with Tippett's construction by blocks, the opposition of dynamic music, with flamboyant orchestral effects, and quieter music concerned with the release of that energy, in a mood that Tippett has called the 'windless night sky and the tidal wave below' (Morris 1999: 461).

Clearly some of the spatial terms found in talk about music are figurative, and understood to be so. Some of them seem, however, to be what Roger Scruton called 'indispensable metaphors'. Even if music is purely of time we use spatial terms in talking about it, and if Scruton is right, we have no choice. Scruton speaks of a 'tonal space', woven together from rhythm, melody and harmony, a space in which *movement* is heard. He indicates something important about musical experience when he writes for example: 'The distribution of pitches in melody is [...] a conquest of tonal space, a movement from and towards' (Scruton 1997: 50). The word 'distribution' rather than 'sequence' suggests the area covered rather than the order in time of various pitches. The experience of melody in traditional western music – Scruton's 'from and towards' – involves a sense of a starting point to which the music will eventually return: having moved away, we are drawn back and would be frustrated if we but were left 'suspended'. A melody may 'soar', but we expect it to land again. The whole movement maps out a space – up or down, away from and back – a more or less closed space, incidentally.[10]

Spatial terms may be indispensable in talk about music, even if they are not indispensable in talk about the actual physical sounds we hear. But to talk simply about sounds or a sequence of sounds is not to talk about music at all. If we conceive of a melody simply as a sequence of sounds at various pitches it is difficult to imagine how we hear a melody as a movement.[11] We might ask what actually moves when a melody 'moves up and down' or 'from and towards'. We do not actually *hear* in one sound after another any 'thing' that moves at all, though we hear a sequence of pitched sounds. In tonal music in the western tradition a note is 'going' somewhere: it is apprehended inside what Scruton calls a musical 'field of force', and the same actual, acoustically identical pitch will be apprehended differently in a different context, as 'moving', leading the listener somewhere else. As an example we could consider the 'leading note' in the western scale: a note one semi-tone below the tonic or 'home' note. The same pitch can become the tonal centre in a piece of music in its key, when it becomes 'home', with a note at a different pitch – vibrating with a different frequency – having the properties of a leading note.

In his book *Individuals*, Strawson produced powerful arguments to the effect that there can be no purely auditory space. The second chapter, 'Sounds', sets up a remarkable thought experiment about a world of pure sound, and reaches the conclusion: 'A purely auditory concept of space [...] is an impossibility' (Strawson 1959: 65–66). Strawson observed that there are 'spatial analogies implicit in our

---

[10] Or 'striated' in Deleuze and Guattari's sense. 'Endless melody', or the construction of melodies in some Indian or Arabic music, may be more 'smooth' or 'nomadic'.

[11] A similar point was made by Bergson, who used melody to clarify the difference between a succession of moments in time and his concept of duration. In *Time and Free Will* he uses music as an analogy for duration: '[...] as happens when we recall the notes of a tune, melting, so to speak, into one another [...] even if these notes succeed one another, yet we perceive them in one another' (Bergson 1913: 100).

ordinary talk about sounds' and continued by noticing the 'persistent and by no means irrational tendency of critics of music and the plastic arts to discuss the formal properties of the works they are criticizing in terms which, in their literal application, belong to each other's domain'. Scruton's 'indispensable metaphor' suggests however that the phenomenon is more deeply rooted than the common transference in discussion of the arts of terms from one domain to another.

There is more at stake in this matter – a more which some of Strawson's thinking may help to illuminate. The sequence of sounds we apprehend as a melody, or part of a melody, becomes, in Strawson's own terms, an 'individual', an 'identifiable' and 'reidentifiable particular'. We recognize it when it recurs, perhaps when it is repeated in a symphony. We recognize it when it is played at a different pitch, when it is transposed. It may even happen that the individual notes are passed from instrument to instrument, and still we hear *the* tune. We may recognize the melody as 'the same tune' when it is varied – at least within limits; and we may recognize it as 'the same' even when it is set to different harmonies. Some of us may recognize a 'musical individual' even when we hear it played upside down, or backwards – as sometimes in Bach's music as well as in Schoenberg's – although this is often tricky for those of us who are not practised musicians. A sequence of notes, if we experience them as a melody, becomes an entity, a kind of 'thing', a re-identifiable particular. This may be one of the reasons for the difficulty of Schoenberg's music: it is harder for the listener to identify and re-identify some of the entities constituted in his sequences of sounds.

There are other musical entities, re-identifiable particulars, apart from melodies: a rhythmic pattern, or a sequence of chords, for example – the 'cadence' which is such an important feature of tonal music is a clear case. In the experience of some listeners there are larger entities, in the 'forms' that they can hear in a whole movement in a symphony or string quartet, and in which they have a sense of unity through the returns, and contrasts, and the movement of those voices that Hegel spoke of. Of course the entities we identify and re-identify may well be ontologically strange. Scruton summarizes the problem as he sees it:

> [If][...] a theory of musical ontology [...] strays into the world where the musical individual is encountered, it is a world of metaphor. If it strays into

> the world of sound, then it can do no more than specify the sound patterns that make the musical experience available. (Scruton 1997: 117)

If we accept this view, then we accept that 'musical individuals' are quasi-objects, quasi-existing in a quasi-space. 'Musical space', in whatever terms it is figured and however differently its quasi-qualities are conceived, remains metaphorical whoever explicitly or implicitly deploys it. Yet although it is true that we are not inside this 'musical space' ourselves, at least some of the 'musical individuals' satisfy all the conditions specified by Strawson for his 'objective particulars': they are re-identifiable, they are distinguished by the thinker from himself, they are distinguished from his own experiences or states of mind, and they are regarded as actual or possible objects of experience (Strawson 1959: 61).

Scruton defends his theory of metaphor by introducing the 'imagination'. Hearing music requires imaginative listening, and accordingly can only be addressed 'metaphorically'. If this is to hold, it requires of the imagination in some of its operations that it be valid for all listeners who apprehend musical entities. It is a shared imagination, independent of any particular listener. Some of the spatial expressions we have encountered are clearly 'imaginative' in one sense. The image offered in Tippett's 'windless night sky and the tidal wave below' seems, however, to be imaginative in a different way from the felt movement of a melody. Tippett's image is not shared by all those who listen to and apprehend his music, while the melodies and other musical entities he composes are – at least sometimes. Re-cognizing those involves cognitive processes of a different kind.

A Kantian model of musical experience might assist in distinguishing the two orders of 'imagination'. Experience is to be understood, in a Kantian way, as more than sensation. It requires the processing of some given. In our case, the given is the sounds we hear. But a sequence of sounds is not yet music. On top of the apprehension of the sounds is a second dimension of experience. The sounds we hear in the first dimension are actual (or imagined). The music is what we experience when we process those sounds with whatever capacities of musical understanding (*Verstand*) we have. We bring to bear something like Kant's *a priori*. In western music, the attempt was made to delineate that *a priori* in the 'rules' of tonality which were intended to operate in a way reminiscent of Kant's categories, as predicates of all possible intelligible musical construction.

The *a priori* we bring to bear on a piece of music cannot be a fixed set of rules, however. Rameau's attempt to associate his harmonic principles with natural principles – laws of nature – implies like Kant's *a priori* a universality that is not valid for music. Schoenberg cast aside the gravitational theory of music and refused to let any one note be 'key', the centre of gravity; such was the rationale for the insistence that all twelve notes of the chromatic scale be heard before any be repeated. Instead of the notes and chords in a piece of music relating to some centre, there is only the relationship between the notes, and the whole 'floats' with no ground to land on, with no home to return to, with no way of indicating in a succession of sounds that 'cadence', or 'falling', that is a marked feature of traditional western music. That may be a factor in the difficulty many feel with his music: it is not always easy to orient oneself in the space he develops. Yet some have learned to listen to Schoenberg and experience music. We can also learn to hear music in the sounds produced by Indian, Arabic, Chinese, Japanese or Javanese musicians for example which do not follow the western division of the octave, despite the claim that it is based on 'natural' acoustics, or of western harmonic or rhythmic principles. In all of these musics there is a movement that can move us. The *a priori*, those principles that allow us to constitute musical individuals, have to be understood as 'here and at present' rather than as universal, as a set of temporary limits; there is always the possibility that we can transcend our current musical understanding (*Verstand*) and overcome the limits implicit in the claim that tonality is based on laws of nature and so prescribes *a priori* forms of musical experience.

The cognitive model of musical apprehension may also explain why a commentator like Schwabe could so misconstrue Chinese music. He wrote:

> Die Chinesen geben vor, sie wären die ersten Erfinder der Musik gewesen, und rühmten sich, sie hätten solche vormals zu der höchsten Vollkommenheit gebracht. Wenn aber das, was sie sagen, wahr seyn sollte: so muß sie nach der Zeit sehr verschlimmert worden seyn; denn itzo ist sie so unvollkommen, daß sie kaum den Namen verdient.[12]

---

[12] Johann Joachim Schwabe (1747–1774), quoted in Giordanetti 2005: 17.

It would seem probable that Schwabe applied his European principles of musical cognition and so found Chinese music lacking.

The cognitive model also suggests that it would be a mistake to assert a singular, universal concept of musical space. The gravitational view is not the only one in the western tradition, and the experience of musical space may be very different in the musics of peoples other than those of the various western traditions.

In his *Kritik der Urteilskraft* Kant argued famously that music has less value than any of the other fine arts because it only plays with sensations. It does not involve concepts and leaves nothing to think about. It is a pleasant, if transitory, experience.[13] Yet it would seem that the experience of music depends on syntheses that have much in common with those Kant analyses for the understanding. The emotional power of music follows for those with the cognitive abilities required for its apprehension. Perhaps this is why the emotional response of many to Schoenberg's music is disorientation, bewilderment, and even anger and frustration. His 'musical ideas' are often organized so densely and make such demands of the listener's cognitive abilities that many of us are unable to make the necessary syntheses. Norman Lebrecht reports that even Mahler had trouble following the score of his Second String Quartet (Lebrecht 1987: 255). Schoenberg's hat is a concrete image and seems to make strong ontological claims. His idea of the manipulation of objects in musical space, however, is not entirely remote from terms used by many others who have reflected on what it means to experience music. Strawson's argument about the impossibility of a purely auditory space may hold for any concept of the *transcendental* reality of musical space, but may not hold for its *empirical* reality. But then, for Kant, the same would be true of the physical space in which we live and breathe.

---

[13] *Kritik der Urteilskraft*, Akademie Ausgabe, V, 328: 'Denn, ob sie [die Musik] zwar durch lauter Empfindungen ohne Begriffe spricht, mithin nicht, wie die Poesie, etwas zum Nachdenken übrig bleiben läßt, so bewegt sie doch das Gemüt mannigfaltiger, und, obgleich bloß vorübergehend, doch inniglicher; ist aber freilich mehr Genuß als Kultur (das Gedankenspiel, das nebenbei erregt wird, ist bloß die Wirkung einer gleichsam mechanischen Assoziation); und hat, durch Vernunft beurteilt, weniger Wert also jede andere der schönen Künste.' Cf. also V, 329: '[...] so hat Musik unter den schönen Künsten sofern den untersten [...] Platz, weil sie bloß mit Empfindungen spielt.'

*References*

Bergson, Henri. *Time and Free Will: an Essay on the Immediate Data of Consciousness*. Transl. F.L. Pogson. London: Allen and Unwin, 1913.

Bowen, Meirion (ed.). *Tippett on Music*. Oxford: Clarendon, 1995.

Bowie, Andrew. 'German Idealism and the arts.' *The Cambridge Companion to German Idealism*. Ed. Karl Ameriks. Cambridge, Cambridge University Press, 2000. 239–257.

Busch, Regina. 'On the Horizontal and Vertical Representation of Musical Ideas and on Musical Space (I).' *Tempo*, New Ser., No. 154 (Sep. 1985): 2–10.

—. 'On the Horizontal and Vertical Representation of Musical Ideas and on Musical Space (II).' Transl. Michael Graubart. *Tempo*, New Ser., No. 156 (Mar. 1986a): 7–15.

—. 'On the Horizontal and Vertical Representation of Musical Ideas and on Musical Space (III).' Transl. Michael Graubart. *Tempo*, New Ser., No. 157 (Jun. 1986b): 21–26.

Frank, Manfred. *Einführung in die frühromantische Ästhetik*. Frankfurt am Main: Suhrkamp, 1989.

Fugmann, Nicole. 'The Gestalt Change of Postmodern Critique: Anselm Kiefer's Spatial Historiography.' *New German Critique* No. 75 (Autumn 1998): 90–108.

Giordanetti, Piero. *Kant und die Musik*. Würzburg : Königshausen & Neumann, 2005.

Guck, Marion A. 'Two Types of Metaphoric Transference.' *Music and Meaning*. Ed. Jenefer Robinson. Ithaca, London: Cornell University Press, 1997. 201–212.

Hegel, G.W.F. *Ästhetik*. Frankfurt am Main: Europäische Verlagsanstalt (nach der 2. Auflage, 1842.), n.d.

Herder, Johann Gottfried. *Sämmtliche Werke*. IV. Teil. Bd. 4, Wien: Geistinger, 1813.

Inwood, Michael (ed.). *Georg Wilhelm Friedrich Hegel: Introductory Lectures on Aesthetics*. Transl. Bernard Bosanquet. London: Penguin, 1993.

Jameson, Fredric. 'Space: Utopianism after the End of Utopia.' *Postmodernism, Or, The Cultural Logic of Late Capitalism*. Durham: Duke UP, 1984. 154–180.

Kant, Immanuel. *Kritik der Urteilskraft*. Akademie Ausgabe, Vol. 5. Berlin: Reimer, 1910ff.

Lebrecht, Norman. *Mahler Remembered.* London: Faber and Faber, 1987.

Morgan, Robert P. 'Musical Time/Musical Space.' *The Language of Images.* Ed. W.J.T. Mitchell. Chicago, London: University of Chicago Press, 1974 (rpt. 1980). 259–270.

Morris, Mark. *A Guide to 20th–Century Composers.* London: Methuen, 1996. [Rept. Pimlico, 1999].

Robinson, Jenefer (ed.). *Music and Meaning.* Ithaca: Cornell University Press, 1997.

Rosen, Charles. *The Classical Style: Haydn, Mozart, Beethoven.* London: Faber and Faber, 1997. (New edition; first published 1971.)

Schopenhauer, Arthur. *Sämtliche Werke.* 5 Vols. Vols. 1 & 2: *Die Welt als Wille und Vorstellung.* Stuttgart: Cotta, 1960–1965.

Scruton, Roger. *The Aesthetics of Music.* Oxford: Clarendon Press, 1997.

Stein, Leo (ed.). *Style and Idea. Selected Writings of Arnold Schoenberg.* Transl. Leo Black. Berkeley, Los Angeles: University of California Press, 1975.

Stravinsky, Igor. *Poetics of Music, in the Form of Six Lessons.* Cambridge Mass, 1970

Strawson. P.F. *Individuals.* London: Mcthuen, 1959.

Zarlino, Gioseffo. *Istitutziono harmoniche.* [Venice 1573]. Ridgewood, New Jersey: Gregg Press, 1966.

# Seeing into Space: The Unconscious and Schematization

Louise Fairfax

In a collection of papers on space, consciousness and art, Kant's aesthetics limned in the *Kritik der Urteilskraft* (*Critique of Judgement*) might provide an obvious starting point. Despite that, I have selected the *Kritik der reinen Vernunft* (*Critique of Pure Reason,* henceforth: KrV) and, in particular, Kant's so-called Copernican Revolution, as my focus. Why this seeming perversion? First, it is in the first *Critique* that Kant outlines the process of schematization, highly relevant to the discussion that will take place here. Even more importantly, it is the Copernican Revolution that specifically marks the end point of the old conceptualization of reality – spatial and otherwise – and, consequently, of the former way art could be understood: as a representation of some absolute reality.

Prior to Kant, it was thought that an object was mirrored in some material way in the brain; the subject merely supplied the sensory receptors and brain as receivers of information from the objective world, and had no active role in the constitution of knowledge. However, this mimetic, *camera obscura* model failed to represent the complexity of the process that actually took place. Kant recognized these inadequacies and argued that if our knowledge must conform to the nature of objects, he didn't see how we could know anything of them apart from segmented experiential sensations. If, on the other hand, objects conform to what we, the subjects, contribute to sensations, then he can understand the possibility of a knowledge that transcends a mere collection of random stimuli.[1] The new emphasis is no longer the object, but the subject: 'unsere[r] Vorstellungsart' (KrV: 33), or our way of seeing things. The objects of experience must

[1] Kant KrV: 32. This is the preface to the second edition, where he explains the revolution that his whole theory as outlined in KrV represents.

necessarily conform to these conceptions. He considered this changed focus from object to subject so momentous he titled it a revolution, and likened it to Copernicus' heliocentric one. This was not an instance of deluded self-aggrandizement or an inappropriate paean to his own ideas. His new approach introduced an entirely new emphasis in epistemology (and ontology).

Even Hume had as datum an objective reality: he assumed a match between representations and 'reality'. Kant questioned this putative truism. As Sebastian Gardner points out, all pre-Copernican systems assume a domain of objects which have a being and constitution of their own – a class of real things (Gardner specifically mentions Berkeley and Hume, 1999: 38).[2] After Kant's Copernican turn, we need to talk of 'ways of seeing', and not 'the way to see'. Once Kant's revolution has been acknowledged, reality can no longer be understood as an absolute or 'Given', to employ the locution of German systems-theorist Niklas Luhmann's English translators. I believe the translators have capitalized this word to foreground its former absolute and authoritative status. I will expand on 'ways of seeing' and the loss of Givens in the section on John Berger.

James Elkins in *The Object Looks Back* adds further dimensions to this Kantian concept. Not only is there no such thing as an object in space demanding to be seen in a particular way: there is no desideratum that it be seen at all. Light can shine on the retina and the resulting electrical impulses travel to the brain with no resulting representation. In fact, Elkins raises the interesting and challenging question of what is omitted in accounts of consciousness when we schematize our representations, and suggests that every act of representation is partial in view of the non-rational aspects involved in our visual schematization. Whether we think of Freud and his unconscious intentions, or of Kant and the role of the *Vermögen* (faculties) in the acquisition of knowledge, this indicates that our intentions, imaginations, general predispositions and desires precede and influence our representations, and therefore underlie or create the

---

[2] Gardner also points out that in pre-Copernican philosophy there is a clear division between 'what is the constitution of reality' (ontology) and 'how do we attain knowledge of reality' (epistemology). The Copernican Revolution blurs this boundary. It is an 'epistemological turn', as epistemological questions and ontological ones are considered from an epistemological angle (40).

visual frame. What enters the plane of consciousness has an entire undisclosed sub-text.

In this article, I wish to throw a Kantian light on ways of seeing as an issue for art and consciousness, particularly as they are articulated by people like Berger and Elkins above. I will enter Kant into dialogue with both Elkins and Berger. I will then extend the discussion to the domain of literature, and demonstrate how Goethe in *Die Leiden des jungen Werther* (*The Sorrows of Young Werther*) in many ways presages the significant Kantian turn within his descriptions of nature and the articulation the characters give to his own new 'interstitial' understanding of reality. Kant gives Goethe's literary adumbrations formal philosophical expression seven years later in his first *Critique* of 1781 and, even more explicitly, six years after that in the preface to his second edition of that *Critique*. My methodology will allow Berger and Elkins to further enhance our understanding of *Werther*, and allow *Werther* to illustrate what later theorists will say about ways of seeing.

## *Desire and the Limits of Seeing*

Sigmund Freud, in what leading evolutionist Stephen Jay Gould amusingly labels 'one of history's least modest pronouncements' (Gould 1987: 1),[3] proclaimed that his theory marked the third and final step (following Copernicus and Darwin) in the humiliating terminus to anthropocentrism: the Ego is not master of its own house (Freud 1974: 326); our unconscious selves rule large segments of our conscious thought and behaviour. He illustrates this in his discussions on hysteria, repression, sublimation and the like – all techniques in which non-conscious thought triggers behaviour. The unconscious self rules silently, invisibly, undetected unless we know how to look. In this context, Elkins talks of a double blindness, pointing out, for example, that in the Freudian slip we are both mistaken, and blind to our mistake. When an idea from the unconscious punctures a hole in our consciousness, it covers not only the rent, but also the act of covering (1996: 217). In fact, Elkins extends the whole Freudian argument, debunking both the idea of conscious determination of our

[3] Gould argues that Freud's list falls short: that there is a fourth unseating of the narcissistic self-aggrandisement of *homo*, and that is 'deep time' – deep geological time, that is.

actions, as Freud did, and also the more significant former truism that vision is objective, that our eyes, at least, are ours to control: 'At first, it appears that nothing could be easier than seeing. We just point our eyes where we want them to go, and gather in whatever there is to see. Nothing could be less in need of explanation' (1996: 11). Rather, he argues, seeing is 'irrational, inconsistent, and undependable. It is immensely troubled, cousin to blindness and sexuality, and caught up in the threads of the unconscious. Our eyes are not ours to command; they roam where they will and then tell us they have only been where we have sent them. No matter how hard we look, we see very little of what we look at'. Seeing is 'entangled in the passions [...] and in pain' (both quotations, cf. Elkins 1996: 11). This gnome is stated in the opening paragraphs; the remainder of the book is an extended argument for, and illustration of, that primary thesis.

To regard vision as being entangled in the passions is to accept in principle Kant's Copernican Revolution which took the conditions for obtaining knowledge away from the object itself and devolved them to the subject. Both Elkins and Kant are in agreement that seeing (Elkins) or representation (Kant) of a spatial image is a great deal more than light striking the retina and travelling to the brain, as a purely objective account would claim. That 'great deal more' for Kant is what brings about his Copernican Revolution – the subjective element in observation. Kant makes us aware that incoming stimuli are acted upon in various ways before representation takes place. *A prioris* of time and space, the categories of understanding and our imagination all play a part in his account. What registers on the plane of consciousness as an object in space only occurs following these processes. This can find a parallel in Elkins's observation that passions, desires and existing knowledge are all important in determining whether or not we see – that is, whether we consciously acknowledge, and are aware of, having seen. 'We imagine that seeing is entirely objective'; however, '[o]ur 'objective' descriptions are permeated, soaked, with our unspoken, unthought desires' (Elkins 1996: 33). Not only does desire direct our gaze towards what it seeks, unconscious desire also filters out much of what is before our eyes. We simply fail to notice things that do not align with desire, he maintains. Our unconscious can protect us from seeing and registering material that it finds too painful or violent, or threateningly sexual. The light rays may go in, but no representation takes place. For

Elkins, it is the unconscious that does this filtering. In Kantian terms, no schematization takes place. Kant would argue that 'reason' has filtered out this potential information. I opine that both thinkers are confluent here, but using different terminology. Kant, of course, hadn't heard of the unconscious; nor does he devote time in this account to desires or passions (although they do enter some of his other considerations; he is not unaware of their influence, but his agenda in the *Critique* leads him to foreground other matters). The *unconscious* operation of reason is implied in his description, in that the unschematized never becomes (conscious) knowledge. It is possible to accommodate Kant into Elkins' more extended and modern understanding, I believe.

Interesting also in the light of Kant and the subjectivity of representation is Elkins' discussion of how knowing more enables us to see more (without any change in the object). Similarly, knowing differently spawns different seeing; and having alternate interests also influences what is seen. Individuals can look differently at different times; one may 'turn on' one's gaze as artist or botanist, selectively and differentially scrutinizing as each commands dominance. One example he gives is of crepuscular rays: the special beams that radiate from the sun as it sets. They arch up and over the vault of the sky, eventually converging in a faint 'antisun' at a point in the east exactly opposite. I have never seen this illusion, as I have never looked with eyes educated to its existence. I have no doubt seen it without seeing. Now I have this knowledge, I am sure I will notice it in the future. Similarly, I can fail to hear the blackbirds singing as I write this, or be entranced by them if I attend. The object has remained constant; my subjective advertency has altered. My brother could sit in my garden and not register that the roses have started blooming or the wisteria smells wonderful. We all have our own examples of this selective attention that results in partial seeing and hearing – or blindness and deafness.

When theorizing about plethoric incoming stimuli, systems-theorist Niklas Luhmann stresses the complexity and enormity of this bombardment, and the need to select from that potentially overwhelming mass in order to prevent system overload. In this regard, he points out that the old notion of consciousness was of a picture of the real world, and emphasizes that it is not enough to replace this with a new notion of consciousness as a reflexive process, as done by Hegel or Fichte. This is too simplistic and fails to embrace complexity. Hegel and Fichte's concept of the symmetry between the complexity of the world and of consciousness prescinded the problem, and the accomplishment, involved in the whole process of consciousness. The asymmetry that Luhmann proposes involves selection and, for him, selection is carried out as a function of meaning, which is itself envisaged in functional terms. Consciousness does not represent the totality of experienced impressions (Luhmann 1990: 30).

Both Kant and Elkins in different ways demonstrate that they are aware of this problem of selection. For Kant, that awareness is evident in his discussion of the need for *a priori* intuitions to impose order on this bombardment, or 'buzzing confusion', as Sebastian Gardner calls it in his discussion of Kant (1999: 72). Kant points out that without the *a priori* imposition of unity and form, I would be nothing more than a separate series of sensations. If I could not comprehend the variety of my representations in one consciousness, my self would be as many-coloured and various as the representations of which I am conscious: 'sonst würde ich ein so vielfarbigen verschiedenes Selbst haben, als ich Vorstellungen habe, deren ich mir bewusst bin' (KrV 112; see particularly sections 16–23). Kant considers our predisposition to see objects as occupying space and situated in time as crucial, and as indicators that his *a prioris* of time and space exist prior to experience, giving it form by which we can comprehend it. In our era that post-dates Einstein and non-Euclidian geometry, it is no longer possible to think of absolute *a prioris* of time or space. I don't believe this justifies our jettisoning Kant, although he was, indeed, mistaken here. I maintain that we can accommodate Kant's notion of *a prioris* into our modern understanding if we regard them as function rather than absolute thing as he intended – if we take into account the tenor of what he was doing as he formulated this theory within the constraints of extant knowledge.

I believe there are two aspects to his concept of *a prioris*, one social and one neurological. On the social side, I opine we can align *a prioris* with social conventions – ways of seeing that operate within a system and influence what and how we see.[4] On the physical plane, I

[4] Michael Friedman (2001) suggests an excellent modification of Kant's notion by employing and extending Henri Poincaré's agreed conventions of mathematics. The *a*

believe we can relate Kant's notion to our biological tendencies, both to our chemical constitution, including hormones, which would relate to Elkins' passions and desires, and to neurological elements – the way we are 'wired' as humans, to employ the locution of leading neurologist, Vilayanur Ramachandran (2003: 9, 23).[5] Ramachandran is another who, like Gould, refers to Freud's analysis of the 'descent of man', debunking ideas of a conscious will making rational decisions for us: 'Your conscious life is, in short, nothing but a post-hoc rationalization of things you really do for other reasons' (Ramachandran 2003: 2). He gives a neurological account of vision, part of which is that once an image is recognized, it is sent to the amygdala, the emotional centre of the brain, which assesses the emotional significance of what has been 'seen' (2003: 8). The Fusiform gyrus is a different part of the brain that enables recognition. He talks of the wiring of the brain to areas associated with creativity, and of cross-wiring and 'excess' wiring, producing different kinds of creativity. We primates have not only a visual cortex; we have thirty visual areas at the back of our brain. These areas work without our conscious awareness (2003: 34). I feel when I am reading this that I am being given a neurological explanation of what Kant posited as *a prioris* providing form and unity prior to conscious representation. This is a fleshing out of unconsciously operating schematism. I don't feel that it contradicts Kant, but that it modernizes and explains him.

---

*prioris* or ways of seeing become agreed on conventions within the system employing that optic as part of its value system (to re-word Friedman using Luhmann terminology).

5 Kant can also be related to Ramachandran in his later discussion on self-consciousness, where he relates it to our awareness of the giving of form and unity to perceptions – the self is only the awareness of my thinking: 'das Ich ist nur das Bewusstsein meines Denkens' (KrV 250). It lacks the conditions of a substance: '[es] fehlt [...] an der notwendigen Bedingung [...] der Substanz' (KrV 250). The subject, the thinking *Ich*, then, is not substance; it is nothing more than the logical qualitative unity of self-consciousness in thought (a function): 'Das Subjekt wird in eine bloße logische qualitative Einheit des Selbstbewusstseins im Denken überhaupt [...] verwandelt' (KrV 250). Nobel laureate, Francis Crick is even blunter: '"You", your joys and your sorrows, your memories and your ambitions, your sense of personal identity and free will, are in fact no more than the behaviour of a vast assembly of nerve cells' (Crick: 1994, 3). Towards the end of his life, Crick concentrated attention on the rarely studied claustrum – its molecular biology, its role in consciousness. This neurobiological understanding of consciousness accords with the tenor of Elkins' argument.

Ramachandran even attempts to posit artistic universals and to seat them in aspects of brain function (2003: 47–50). This latter idea, however, is far from being generally accepted in his field, and at a recent seminar series, 'Frontiers of Consciousness. Chichele lectures 2006', at Oxford University, corralling many of Britain's leading cognitive scientists and neurologists, not one neurological researcher mentioned it.

When Elkins says we are predisposed to find bodies and faces in ambiguous constellations of stimuli, I believe he is making oblique reference to the same aspect, which Ramachandran relates to evolutionary principles and needs of the species. In fact, the same epithet from above is employed by Elkins in his treatment of this topic. He argues that if we noticed every detail, 'the world would turn into a fluttering, buzzing confusion' (1996: 97); 'No seeing sees everything' (1996: 95). Whereas Luhmann deploys the concept of meaning as the determining function for selection, Elkins talks of passion and desire as the sieves used by the unconscious; Ramachandran of 'emotional significance'. I regard these as merely different semantics describing the same process. The crucial concepts are complexity and unconscious selection and processing preceding representation. Elkins is also cognizant of an additional problem of finitizing the complexity of reality in representation, and refers in this context to the pain of finitization in the selective glance: 'every glance hurts by freezing and condensing what's seen into something that it is not' (1996: 29). When I later discuss Werther who agonizes over this very problem, Elkins' observations here will become relevant.

Lastly, it must be mentioned before proceeding, that Kant and Elkins share an understanding whereby subject and object have merging dependence and not discrete spatial realities. For Kant, the object only becomes known by our representations, and is thus dependent on the subject for the particular kind of knowledge we gain of it. Meanwhile, subjects become aware of their own subjectness and the unity of their consciousness through the act or function of apperception. Elkins words it: 'I don't really exist apart from the objects I see'; 'There is no such thing as pure self, or a pure object without the self' (1996: 44). Elkins has a huge yet unstated debt to Kantian thought.

*Seeing as Second-Order Observation*

John Berger in *Ways of Seeing* also expatiates on the theme that vision is not objective. Pre-empting Elkins' first two chapters, he writes pithily: 'To look is an act of choice' (1981: 8). As a direct implication of the Kantian Revolution, he notes that an image is not, in fact, a representation of an object, Y, but 'of how X had seen Y' (10). Artists and artistic photographers are very aware that they are presenting their particular way of seeing and not an object itself. Berger argues that this was not always the case; that with a few exceptions (the masters who transcended their medium) oil paintings, which reigned as the dominant form of representation in what can be termed the ontological age (Andersen 2003: XI), or era of the 'thing schema',[6] were about presenting objects – or, more to the point, possessions: 'Oil painting, before it was anything else, was a celebration of private property' (Berger 1981: 139). Berger draws a direct continuum from this kind of oil painting to modern advertising: 'It is a mistake to think of publicity supplanting the visual art of post-Renaissance Europe; it is the last moribund form of that art' (139). Berger's primary point has to do with the philosophy behind this object-orientation and the narcotic and blinding desires it engenders. What I wish to emphasize is his particular presentation of the theme that images are not primarily of objects, even when they purport to be: they are of ways of seeing objects. The presentation of an image contains a way of seeing and an unwritten philosophical stance, which I would relate to the *a priori* aspect of Kant's schemata (which he considers to be connected to both the experiential and to *a priori* categories). In the ontological age, the object was assumed as an unequivocal Given, and that assumption is reflected in the object-oriented presentation of that age. In a post-ontological era, the particular way of seeing is foregrounded in the most perceptive art. The subjective constitution of reality is part of the image; implicitly, the Kantian revolution is acknowledged. Each depiction, in fact, in Luhmann's terms, is a second-order observation; that is, paintings or photographs are observations that entail a particular observer position, and cannot be seen as merely first-order

[6] Before modernity, the 'thing schema' was universally valid: 'war das Dingschema universal gültig' (Luhmann 1984: 98). He directly ties in the de-ontologization of modernity with the loss of this thing schema (Dingschema): 'Dies führt zu einer radikalen De-Ontologisierung der Perspektive auf Gegenstände schlechthin' (243).

observations of things as Givens, demanding a certain perspective. As Berger words it: 'the visible no longer presented itself to man in order to be seen' (1981: 18). As Luhmann will later word it: 'knowledge can no longer be tied to some Given that every reasonable person is capable of discovering' (1990: 61).[7] Niels Andersen, writing about this topic notes: 'Second order observation is aware that the world is not asking to be observed in any particular way. Second-order observers perceive the world as poly-contextual, as dependent on the distinction shaped by observation. The guiding distinction steers the observation and frames the choice of other distinctions' (Andersen 2003: 95), and he talks of the collapse of the 'innocence of the empirical' – 'when we can no longer pretend that 'the object out there' discloses how it wants to be observed' (2003: XV). This is the collapse that underpins Berger's thesis.

Once more I emphasize that, philosophically, this can only occur in a post-Copernican era, where Kant has dismantled the idea of an unequivocal Given and stressed the importance of the subject's way of seeing. These new ways of seeing, as adumbrated above, were tied to the specific breakdown of the old absolutist way of seeing, to the acknowledgement of perspectives and an assortment of valid subjective viewpoints.

Berger points out that in the old object-oriented (ontological) era, landscapes were not popular. They did not represent objects and possessions, and sky was not even a tangible thing. Before the modern ecological movement, nature was merely 'the arena in which capitalism and social life [...] had its being' (Berger 1981: 105). It did not reflect hesternal desires and orientations, and thus did not find direct representation. He notes: 'The first pure landscapes – painted in Holland in the seventeenth century – answered no direct social need. (As a result Ruysdael starved and Hobbema had to give up)'. To paint landscape thus represented an alternative way of seeing which already marked a turn away from the object. The innovations of the landscapists 'led progressively away from the substantial and tangible towards the indeterminate and intangible'. Furthermore, 'each time the tradition of oil painting was significantly modified, the first initiative

---

[7] Luhmann calls the notion of 'correct concepts' or a pre-given world 'naïve' (1990: 21). A single, universal and binding meaning attached to an object was replaced by the notion of a multiplicity of meanings made available by different perspectives.

came from landscape painting' (Berger 1981: 105). He argues in this regard that Gainsborough's painting of Mr and Mrs Andrews is not a picture of a Rousseauan return to nature! Mr and Mrs Andrews are, above all, owners of the land depicted: 'their proprietary attitude towards what surrounds them is visible in their stance and their expressions' (Berger 1981: 107).

In the field of literature, Goethe's Werther is one who saw things differently, who did not see nature with a 'thing' orientation or with reference to something outside itself, and who introduced a whole new way of seeing – of nature, of love, of rationality. Part of what makes Goethe so special is this new way of seeing – the fact that he does not overlook what most people ignore, but trains his lens on those things with precision. Goethe's eponymous hero and alter ego, Werther, is part of a new subjectivity that heralds the breakdown of the old ontology. In the section that follows, I will discuss his new vision as it pertains to one aspect of a much greater whole: his new way of seeing nature.

*'Werther': a New Vision of Nature*

Goethe was one of the first, not only to transvalorize nature, but to notice it at all. Elkins discusses how his seeing a certain kind of moth on tree trunks is seeing what others don't see, what they overlook with their different interests and desires. Werther's account of the minute detail amongst the blades of grass in an early section of the book represents the same phenomenon: he is seeing what others had overlooked, heralding a new set of values and a new orientation. He is not interested here in things or possessions. He loses the self in the world of the minutiae of the brumous valley, noticing the previously prescinded teeming world of insects amongst the ribbons of green:

> Wenn das liebe Tal um mich dampft, und die hohe Sonne an der Oberfläche der undurchdringlichen Finsternis meines Waldes ruht, und nur einzelne Strahlen sich in das innere Heiligtum stehlen, ich dann im hohen Grase am fallenden Bache liege, und näher an der Erde tausend mannigfaltige Gräschen mir merkwürdig werden; wenn ich das Wimmeln der kleinen Welt zwischen Halmen, die unzähligen unergründlichen Gestalten der Würmchen, der Mückchen näher an meinem Herzen fühle, [...]; dann sehne ich mich.[8]

[8] Goethe 1948–64: 9. This Hamburg edition of Goethe will henceforth be referred to as HA.

Werther notices; and as our discussion of Kant and Elkins has shown, new seeing does not involve new *objects*, but rather a new way of looking. Werther's seeing is not just a turning from the acquisitive world of possessions, which it also is, but denotes a new recognition of the rights of nature, of the complexity of nature, and of existence *per se*. He, unlike the majority of his peers, does not see nature in relationship to some external reference such as the judgement of God, but as part of our shared world. His reference is not reference to an absolute authority and a single correct way of seeing, but to the self. Because, for him, nature is not a sign of God's damnation for man's sin, and thus abjured or at best ignored, and because insects do not for him represent the nadir on some scale of absolute rationality, Werther is able to see, really see, and attend to detail. Nature, for Werther, is not an expression of anything other than itself and is legitimized by virtue of its own unfolding, its own dynamic being.

It is surprising how seemingly inconsequential, tiny comments can indicate that a new way of seeing has taken place. Having described the busy world of the grassy jungle, Werther turns his attention to the glories of spring and comments that one could wish to be a may-beetle in order to waft in a sea of wonderful scents: 'man möchte zum Maienkäfer werden, um in dem Meer von Wohlgerüchen herumschweben [zu können]' (*Werther*, HA 6: 8). This signifies a very deliberate turning from the old ways of seeing, where the human being was the *ne plus ultra* of God's creation, where reason was equated with being, where vision was the acme of the senses (it being the most reasonable and least corporeal, it was felt) and where the animal world in general was spurned for its physicality and feculence and where the insect world in particular was beneath contemplation. Not only does Werther contemplate – he specifically imagines that being an insect and floating on perfumed breezes would be pleasurable. This is a totally new way of envisaging nature that sees far more than his coevals saw. Werther asserts his difference. Descartes, like Plato before him, had wished to be pure mind in order to discern truth.[9] Werther imaginatively abandons the mind to be pure sensation in arthropodan casing.

---

[9] For example, he writes in the third meditation: 'Ie fermeray maintenant les yeux, ie boucheray mes oreilles [...]' (Descartes 1982: 27).

Important also is the new attitude to nature itself that Werther's statement implies. It is an attitude whereby nature is conceived as positive, and not because it reflects the order of reason. It is accepted on its own grounds, simply because it exists. He does not spurn its unformed chaos or exclude it as ungodly disorder. Werther's attitude to nature is an important aspect of the emergence of a new way of understanding the world – ultimately a new perspectival paradigm of thought.

The old ontological paradigm ultimately viewed the 'good' parts of nature as evincing the order of reason. The absolute God was conceived of not only as a God of reason, but also of order. The act of creation was accordingly envisaged not as a bursting of life force in all its complexity, but as the imposing of order (good) onto chaos (bad). Dualism and ontology underlie this paradigm that posits reason as an absolute ideal. And Werther rejects that orientation. He shares neither this hope of order nor this desire for a tamed, reduced and excluding nature – one that shunned the disorderly and wild that he recognised to be present and legitimate. Like the dissentient passions, nature, for Werther, is to be admitted as it is – merely as and for itself.

Even before he published *Werther*, in 1772, Goethe had oppugned the excluding and divisive paradigm of his day in a review of Sulzer's book on beauty. There Goethe argues by rhetorical question: Are not raging storms, floods, rains of fire, subterranean glow and death just as legitimate signs of eternal life as the magnificent setting sun?[10] In other words, is not wild, disorderly and even frightening fury just as legitimately wonderful as what you think of as traditionally beautiful? What we see in nature, he argues, is a force – one that can, indeed, devour, but is legitimate none the less. Beautiful and ugly, good and evil all share an equal right to exist. Existence is its own legitimation.[11] The same attitudes emerge in a later poem, *Das Göttliche*, of 1783.[12] Nature in this poem does not reflect the sentimental, prettified nature of Sulzer. It is not just wild; it is uncaring, impersonal, heedless of humans, and yet embraced by the

---

[10] 'Sind die wütenden Stürme, Wasserfluten, Feuerregen, unterirdische Glut und Tod in allen Elementen nicht ebenso wahre Zeugen ihres ewigen Lebens als die herrlich aufgehende Sonne?' (Goethe, Johann, *Die schönen Künste in ihrem Ursprung, ihrer wahren Natur und besten Anwendung, betrachtet von J.G. Sulzer*, HA 12: 17).

[11] 'alles mit gleichem Rechte nebeneinander existierend' (Goethe, HA 12: 18).

[12] Poem: HA 1, 147–8.

poet for itself. Werther also welcomes such an unfeeling, impersonal and arbitrary nature for its own sake, and not for its reference back to a God of reason or order or beauty – or anything. Werther advocates a new way of seeing – that of self-reference, and self-legitimation.

This is particularly clear in the segment where he rages with his own ferocity in a sustained metaphor against timid denizens living near a river, making pusillanimous attempts to curb its mighty violence by building little damlets, ditches and Lego-bulwarks (my expression, not Goethe's: he employs *Dämmern* und *Ableiten*) for their riparian tulips and *Gartenhäuschen* (little garden houses – the size no doubt meant to match their stature) in their attempts to emasculate the power of nature rather than respect it (Goethe, *Werther,* HA 6: 16). This verbal picture Werther paints is intended as a picture symbolizing all who need to have rules rather than let nature and their passions and their creative genius flow. His age was an age of rules and reference to higher authority. Once more, he is demonstrating a very different way of seeing that steps beyond the bounds of his era, providing new horizons and a new framework for those who cared to follow his new vision.

It is in Werther's love of nature, in his wonder and delight, that he encounters the problem of representation, expressed later by Kant. This is also a problem articulated by Elkins when he referred to the pain entailed in freezing a moment and selecting from the complexity of the whole to capture a mere segment (Elkins 1996: 29). Werther cannot represent the boundless complexity of nature in the finite terms of plastic art. He cannot reduce it, yet he can take it in, feel it, and respond to its limitlessness.[13] This passage not only introduces us to the general problem of representation that any artist faces, and which is a problem central to Werther himself. He is extremely sensitive to complexity, as his response to nature indicates, and cannot deal with the loss of complexity to reduce it into a two dimensional frame. Like Elkins, he feels pain at that compulsory selection. Soon we will see he cannot and will not reduce the complexity of Lotte, his beloved, into a portrait of her, and settles instead for a silhouette. A portrait of Lotte

---

[13] 'Ich bin so glücklich, mein Bestes, so ganz in dem Gefühle von ruhigem Dasein versunken, daβ meine Kunst darunter leidet. Ich könnte jetzt nicht zeichnen, nicht einen Streich, und bin nie ein gröβerer Maler gewesen als in diesen Augenblicken' (Goethe, *Werther*, HA 6: 9). For an interesting treatment of this statement within the context of the aporia between artistic vision and expression, cf. Graham 1962: 3–24.

would have highlighted the object itself – as Berger's discussion makes clear. Portraiture, by detailing the features of the object and minimalizing those of the background or environment, reflects an ontological orientation, proclaiming the primacy of the 'thing' that dominates the picture. A silhouette, on the other hand, throws into relief the very complexity that a portrait elides – the environment. It is an indirect reference to the light itself that produces it, and to the complexity that is light. In that the ontological features of the object vanish, the object itself has no given status, but is highly subjective. It in fact represents a Kantian Copernican turn to the subject several years before Kant has specifically indited it. The object lies somewhere between the painter and infinity, and represents the boundary between form and infinity or radical complexity. Thus Werther's failure to represent is highly significant. As well as indicating the artistic problem of representation it heralds the rejection of ontological orientation, a turning from the object to the subject, and an unwillingness to reduce complexity in any of its forms.

The final aspect I wish to touch on in a discussion of Werther's new way of seeing nature (as indicative of his new way of seeing *per se*) is to take up a point emphasised by Elkins, that '[t]here is no such thing as pure self, or a pure object without the self' (Elkins 1996: 44). As the title of his book announces, he does not envisage seeing as a one-way process on the part of the subject, taking from the object and leaving it unaltered. Rather, 'seeing alters the thing that is seen and transforms the seer' (11–12). Because he understands it in this way, he intones: 'Seeing is metamorphosis, not mechanism' (12). For Elkins, subject and object have a reciprocal relationship that sees boundaries merge. Neither stands alone and 'pure'. Kant's Copernican revolution also acknowledges that the being of the object is dependent on the subject's representation, and that the subject is aware of itself as subject because of this process of conscious awareness giving unity to representation (cf. esp. section 16 of the KrV). In other words, for Kant there is interdependence between subject and object. This same fusion can be seen before Kant in Werther, for he, too, dislimns the boundaries between subject and object. This is particularly the case once he becomes melancholic. We see from the entry of August 18, three days after he has seriously considered suicide, that nature has become a projection of his self – and this self is now depressed and tenebrous. Now nature becomes unbearable torment, a torturer, the source of his misery: 'die Quelle seines Elendes' (*Werther* HA 6: 51). His capacity to structure objective connections with his environment collapses, and nature becomes a grave as he projects his own morose and moribund depression onto the outer world. It becomes callous and destructive in its corrosive power.[14] Werther is here what Goethe himself calls 'mad'. The scene, however, is merely an exaggeration of what has preceded; it is not an entirely new phenomenon. The melding of subject into object illustrates, I believe, that for Goethe along with Kant and Elkins, and implicitly Berger, the object does not stand alone, independent of our representations.

This article has brought Kant and Goethe into dialogue with modern understandings of consciousness, such as those provided not only by Elkins, Berger, and neurologist Ramachandran. By foregrounding these relationships, I believe I have provided insight into how we can adapt Kant to modern ideas, and how Kant and Goethe can enrich our modern conceptions by fleshing out some of their assumptions.

*References*

Andersen, Niels. *Discursive Analytical Strategies*. Bristol: The Policy Press, 2003.

Berger, John. *Ways of Seeing*. Harmondsworth: BBC and Penguin, 1972 (1981 reprint).

Elkins, James. *The Object Stares Back: on the Nature of Seeing*. New York: Simon & Schuster, 1996.

Crick, Francis. *The Astonishing Hypothesis: The Scientific Search for the Soul*. New York: Scribner, 1994.

Descartes, Rene. *Méditations traduction français, Oeuvres de Descartes*. Vol. IX–I. Charles Adams and Paul Tannery (eds). Paris: Libraire philosophique J. Vrin, 1982.

Freud, Sigmund. *Introductory Lectures on Psycho-Analysis*. Harmondsworth: Penguin, 1974 (A Course of Twenty-Eight Lectures Delivered at the University of Vienna). *Gesammelte Werke. (Werke aus den Jahren 1917–18*). Vol. XI. London: Imago, 1947.

[14] 'Ich sehe nichts, als ein ewig verschlingendes, ewig wiederkäuendes Ungeheuer' (*Werther* HA 6: 53).

Friedman, Michael. *Dynamics of Reason.* Stanford: CSLI Publications, 2001.

Gardner, Sebastian. *Kant and the Critique of Pure Reason.* London: Routledge, 1999.

Goethe, Johann. *Die Leiden des jungen Werther* (First, 1774). *Goethes Werke.* Vol. VI. Erich Trunz et al (eds). Hamburg: Christian Wegner, 1948–64.

Gould, Stephen Jay. *Time's Arrow, Time's Cycle: Myth and Metaphor in the Discovery of Geological Time.* Harmondsworth: Penguin, 1987.

Graham, Ilse. 'Minds without Medium. Reflections on *Emilia Galotti* and *Werthers Leiden.*' *Euphorion* 56 (1962): 3–24.

Kant, Immanuel. *Kritik der reinen Vernunft. Ungekürzte Sonderausgabe zum Kantjahr 2004.* Wiesbaden: Fourier, 2003. (First, 1781; second 1787).

—. *Kritik der Urteilskraft. Ungekürzte Sonderausgabe zum Kantjahr 2004.* Wiesbaden: Fourier, 2003. (Second edition, 1793. First, 1790).

Luhmann, Niklas. *Essays on Self-Reference.* New York: Columbia University Press, 1990.

—. *Soziale Systeme: Grundriß einer allgemeinen Theorie.* Frankfurt a. M.: Suhrkamp, 1984.

Ramachandran, Vilayanur. *The Emerging Mind: The Reith Lectures 2003.* London: Profile, 2003.

# REPRESENTATION, CONSCIOUSNESS, IMAGINATION

# Bodies and Stairs: Modernist Theatrical Space and Consciousness

Paul Monaghan

Western notions of human consciousness have been embroiled in the mind-body dichotomy and consequently intertwined with the development of Western metaphysics (cf. Lakoff and Johnson 1999: 3–8). The so-called 'hard question' of consciousness studies concerns the way in which subjective human experience is generated by and related to the processing of sensory data received from an 'objective' world. 'What is it like to be human?' is the question faced by those who probe the enigma of both consciousness and metaphysics (cf. Chalmers 1995, Baars 1997, Blackmore 2005), and for many at the turn into the twentieth century, it was not philosophical or scientific enquiry but art that was best able to examine these enigmas. Indeed, as Michael Bell argues, 'a central philosophical feature of Modernism [...] is its claim for literature itself as a supreme and irreplaceable form of understanding' (Bell 1999: 29). But as Science and its corresponding artistic forms, 'naturalism' and 'realism' were seen increasingly to be incapable of expressing the industrialized, alienated and fragmented realities of modern life, the Modernist search to understand 'what it is to be human' turned away from a realism that corresponded to the visible external world ('correspondence realism'), and towards a 'realism' that was sifted through or even created by the artist's consciousness ('coherence realism') (cf. Grant 1970: 9). Coherence realism celebrates the subjective imagination; there, reality is not so much found but *made*. In this sense, the coherence theory of realism 'is the consciousness of literature: its self-awareness, its realization of its own ontological status' (Grant 1970: 53). According to this view, the so-called 'reality' of daily life is 'nothing except as led into the arena of significance by a responsive consciousness' (1970: 53). Reality here is seen as *interactional*. Or, as Merleau-Ponty asks: 'How would the painter or poet express anything other than his encounter with the world?' (Merleau-Ponty in Kearney 1994: 82). Modernist art, as Childs argues, 'is associated with attempts to render human subjectivity in ways more real than realism' (Childs 1999: 3).

Modernist theatre is of particular interest here for a number of reasons. From Plato to Descartes and into the present day (cf. Baars 1999), proscenium arch theatre has frequently been used as a metaphor and working model for the study of human consciousness. Plato's 'cave' allegory and Descartes' so-called 'theatre in the brain' foreshadow, as Wiles argues, the 'retreat of the actor into a [proscenium arch] frame', for Cartesian dualism, his notorious separation of the Mind, or *res cogitans,* from the body, *res extensa*, and his postulation of the central, processing pineal gland, made theatre primarily an ocular activity in which a detached, thinking and essentially passive spectator observed an active, embodied actor (cf. Wiles 2003: 4–7). Baars' use of the metaphor is certainly reliant on a (somewhat simplistic) notion of theatre dominated by the proscenium arch. Others, such as Daniel Dennett and Susan Blackmore, dismiss the ultimate usefulness of the theatre metaphor while admitting that it continues to accurately describe the *sensation* of consciousness (cf. Dennett 1993; Blackmore 2005: 13, 44–5). And certainly at the turn into the twentieth century, what we see is precisely a Modernist reaction against the proscenium arch and a move towards the use of a single, unified chamber which both performer and spectator share. However appropriate or not the theatre metaphor is for the study and understanding of consciousness, in the actual theatre we have an experimental space, the world in a crystallized and miniature form, in which to study it. The aural, visual, spatial, and kinetic performance textures we experience and/or observe in theatrical space cause a range of responses in the consciousness of both performer and spectator, with the result that we make connections, interpret, imagine, 'see' things other than what is in the actual visual field before our eyes, and so on. And one can also by-pass the acting event altogether, with all its 'phenomenal distraction', as Bert States calls it, by resorting to private reading, where 'everything is susceptible to envisionment […] however fantastic or surreal the image, it is all real in the sense of its springing to an imagined actuality' (States 1985: 28). Depending on one's bent, this virtual realm is seen as a separate, mind-independent, metaphysical space given access to by theatre, or simply synonymous with the subjective space of our imaginations. Consciousness, metaphysics and theatre, then, appear to be intimately linked, whether one views the relationship from the perspective of what the study of consciousness and metaphysics can contribute to an understanding of theatre, or vice versa. The separation, or alternatively the mutual embeddedness of mind and body, of subject and object, are the substance of this relationship from whichever vantage point one is looking.

In this paper I examine the relationship between theatre, consciousness and metaphysics at the turn into the twentieth century, a time when shifting metaphysical positions both explicitly and implicitly drove Modernist – here specifically Symbolist[1] and Expressionist – theatrical experimentation. I argue that this shifting ground is especially manifest in the way Modernists dealt with that fundament of the theatrical medium, the arrangement of bodies and objects in space. My examination will focus on the use of a special kind of object in theatrical space, namely flights of stairs and multiple-level stage floors, as well as the kinds of bodies that were located on those stairs. Symbolist and Expressionist lighting is also examined, because particular lighting effects were used during this period to complement the use of stairs to create a sense of ascent and descent, of verticality, volume and suspension. I trace a trajectory that begins with two-dimensional Symbolist paintings of stairs, and the Symbolists' 'theatre of the mind' that sought, ultimately, to maintain a neo-Platonic and Cartesian separation of mind and body and hence eradicate the body from both consciousness and theatrical space. The pioneering work of Adolphe Appia and his 'rhythmic spaces' in which the body is regulated by music then takes us through to Expressionism's embodied 'phenomenal' theatre and spaces in which the frequent use of actual stairs is tied to theories of ecstatic physical performance. This shift, a move from proto-Modernist to a more fully Modernist theatre, forms a 'play within a play', so to speak, inside a broader shift from Kant's 'Copernican Revolution' to the (still contested) concept of 'embodied consciousness' advanced by many today, including, for example, Lakoff and Johnson (1999).

---

[1] Whether or not Symbolism is seen as 'Modernist' depends, I think, on whether you are looking at it forwards from 1870 or backwards from 1930, but either way Symbolism was a crucial turning away from Romanticism and towards Modernism.

*The Symbolist Staircase: Bodies Ascending*

The following is a description of a haunting poster that appeared in the streets of Paris early in 1892:

> [The poster] depicted three female figures, one of them nude and sunk into the mire of daily life, slime dripping from her finger tips. The remaining two ascend a celestial staircase. Of these, one is darkly dressed and occupies the middle ground. She offers a lily to a near-transparent figure higher on the stairs who has left life's pollution far behind. This latter figure represents pure Idealism […]. Masses of clouds and stars swarm about the mountain peaks at the top of the stairs. (Pincus-Witten 1976: 102; see figure 8)

The poster, by Carlos Schwabe, announced the first Salon of the Order of the Rosy-Cross and the Grail and the Temple, the first of a series of mystical, Idealist exhibitions and performances in Paris from 1892 till 1897, under the leadership of Josephin Peladan, known as *le Sâr* (an ancient Assyrian word for magus or king). The theme of the poster is 'Initiation': by ascending the stairs of this quasi-secret society and the Idealist art that it practiced, the initiate would gain access to higher truths. The higher one went up the staircase, the poster suggests, the less corporeal would be our bodies. The stairs in Schwab's poster also evoke the ascent to the sunlight in Plato's 'cave' allegory. Those unfortunate human beings confined inside the cave, watching the shadow play of daily life on the cave wall, are symbolized by the 'nude […] sunk into the mire of daily life', and the ethereal figure further up the stairs is on her way to a pure perception of Truth outside the cave. The aim of these exhibitions and performances was partly to represent Symbolist ideals, but perhaps more importantly to actually help bring about the (apparently immanent) transformation of the mud of daily life into the transcendent sunshine of neo-Platonic truth mixed with occult catholicism.

For the Symbolists, stairs like the ones in Schwabe's poster symbolized the transitional status of human existence. The staircase is a particularly potent symbol in this regard, in that it suggests humanity's potential access to higher truths as well as the possibility that these higher truths might influence material life. In the context of 'Symbolism' as a movement of the late nineteenth – early twentieth centuries, the term 'symbol' has metaphysical connotations. According to the Russian Symbolist poet and playwright, Vyacheslav Ivanov, the symbol has the power to take us from *realia* or 'the real' (in the ordinary sense) to *realiora*, the 'more real' (West 1970: 57). Fellow Russian Symbolist Andre Bely asserted that the symbol also had the potential to 'render the immaterial material', to manifest the 'more real' in daily life (Morrison 2002: 3; West 1970: 87). The symbol reveals or suggests, but does not directly name, the hidden essence and meaning behind material life. As Henri de Regnier, a follower of Mallarmé famously asserted in 1900, 'a symbol is a kind of comparison between the abstract and the concrete in which one of the terms of the comparison is only suggested'.[2] Symbolist metaphysics explicitly rejected the concrete and phenomenal in favour of the abstract and noumenal. Human existence, or at least material existence in the form it had taken by the end of the nineteenth century, was seen to be a painful alienation from the divine Idea, and the ultimate aim of Symbolist art was precisely to transform this material reality into a spiritual realm of (largely Christianized) Platonic forms.[3]

A number of Symbolist paintings of the period explored the alienated space of human life by means of the staircase. A new edition of the Holy Bible, published in 1886, featured Gustave Doré's *Jacob's Ladder*, a depiction of the story in which Jacob dreams of a ladder that reaches up from earth to heaven. In the painting, a long, ethereal staircase dominates the painting, and bears a considerable likeness to Schwabe's poster some six years later. In Georges Rochegrosse's *Madness of Nebuchadnezzar*, 1886, set in the ancient Babylonian past (Bohrer 2003: 260, fig. 67),[4] stairs depict a downward 'fall'. At the top of the stairs, light shines on a group of onlookers who stare down a staircase at the king sprawled at the bottom in the semi-darkness. The cause of his madness, a semi-transparent spectre of an angel with sword drawn standing on the King's head, is also visible. In another

---

[2] Henri de Regnier, 'Poètes d'aujourd'ui et poésie de demain.' *Mercure de France* 35 (1900): 342 (cited in Balakian 1982: 27).

[3] The tendency to regard the phenomena of human life as alienated phenomenal objectifications of the divine, or noumenal, realm, with, however, both the desire and the possibility of reunion with divinity, had been strong in German Idealist philosophy throughout the eighteenth and nineteenth centuries. See, for example, the views of Wilhelm Ferdinand Solger (1780–1819), *Philosophy of Art*, cited in Szondi 2002: 24.

[4] The next painting, *The Fall of Babylon*, is reproduced as figure 6 on page 262 of this work. It is possible that the setting of paintings such as these was influenced by the vaulted domes of the Assyro-Chaldean Gallery in the Louvre at this time, and by the heightened pace of archaeological discoveries in the late nineteenth century in Europe.

painting by Rochegrosse, *Fall of Babylon*, 1891, a huge set of stairs dominates the centre of the canvas. At the top of the stairs there is a throne; at the bottom and strewn around the floor area are naked and semi-naked women lying in their own disgrace. The room they are in is vaulted and monumental, a picture of a once proud but now all too human palace. Other Symbolist painters, such as Gustave Moreau, had expressed interest in palaces and ancient Jewish and Egyptian mythology in which small sets of steps are centrally placed.[5] Some spaces in Symbolist painting that allowed and encouraged one to dream of transcending the material world were associated with water. In these kinds of spaces, says Jullian, one might see 'those who wished so fervently to become spirits that they soon became nothing more than shadows' (Jullian 1973: 21).

An extreme simplicity and focus on painting (and poetry) precluded, as far as I am able to ascertain, the use of actual staircases and multi-level floors, but Symbolist stage design took from Symbolist painting its aesthetic, and from Symbolist metaphysics its aims. So important was painting for the Symbolist theatre that Paul Fort, at the *Théâtre d'Art,* announced in Jan 1891 that they would be closing each evening with static *mise en scène* of a Symbolist painting as a *tableau vivant* (Beacham 1987: 5). They do not seem to have put this idea into practice, but a typical Symbolist design was an ethereal space similar to that of the paintings. In the 1891 production at the *Théâtre d'Art* of *Les Filles aux mains coupée* by Pierre Quillard, for example, which Deak asserts 'marked the invention of symbolist stage design', a thin, transparent scrim across the width of the stage, placed just upstage of the footlights, together with the soft lighting, bare stage and a backdrop painted in rich colours by Paul Sérusier, would have set the stage space apart, making it a clarified space suggestive of otherworldliness.[6] The sense of vertical aspiration symbolized by the painted stairs was, in any case, present in Symbolist theatre despite the lack of actual staircases. Deak points out that much Symbolist theatre involved notions of initiation – a notion implying upward movement – and that these plays fell into two categories: 'inner development, in which individual stages of this development are part of the dramatic plot', and liminal or 'static drama' in which 'the hero is suspended

---

[5] For example, 'Salomé Dancing' 1874–76, reproduced in Jullian 1973, fig. 89, and 'Salomé' 1876 (Jullian 1973: fig. 102).

[6] This and the following quotations are from Deak 1993: 142–4.

between two states'. The plays of Peladan generally fall into the first category, and those of Maeterlinck fall into the second (Deak 1993: 127–8). Moreover a major influence on the Symbolists as a whole was Richard Wagner's intensely spiritual music and staging. A photo of the 1882 production of his *Parsifal*, at the Festspielhaus Bayreuth Bildarchiv (reproduced in Baugh 2005: 22, fig. 6), shows the religious monumentality of the stage area with a strong sense of volume and connection to the heavens (in both an upward and downward direction). Light seems to shine on the semi-circular group of performers from the huge dome above them.

The absence of actual stairs on the Symbolist stage is tied to their theatrical aims, which in turn is driven by Symbolist metaphysics. For Peladan and fellow Symbolists, poetry was the primary form of art,[7] and their attitude to the insistent corporeality of live theatre was notoriously ambiguous. Wiles' description of Cartesian theatrical space as 'ocular space' which 'does not submit to any embodied immersion in space' (Wiles 2003: 7), is particularly appropriate as a description of Symbolist theatre. Maeterlinck complained in 1890 of the 'disgust which all artists feel as the curtain rises' and calls the theatre of his day 'the prison of dream – the gaol of art' (McGuinness 2000: 91). The 'malaise' they felt in the theatre was caused not only by the fact they judged the theatre of the day to be base, but because for the Symbolists there was a fundamental antipathy between the 'sullying corporeality' of theatre on the one hand, and the extra-physical aspirations of the symbol on the other. Maeterlinck claimed that '[a]ll masterpieces are symbols, and the symbol can never sustain the active presence of the human being' (cf. McGuinness 2000: 94). Theatrical performance, said Maeterlinck, 'puts things back exactly where they were before the arrival of the poet' (McGuinness 2000: 97).[8] Private reading, Mallarme's 'theatre of the mind' or conceptual theatre is therefore to be preferred.

Despite the difficulty they had with the body, there were over ninety new Symbolist productions in last decade of the nineteenth century (McGuinness 2000: 101). Their difficulty was to to remove

[7] Baudelaire wrote, 'poetry is the most real thing we have, what is only made completely real in another world': cf. Grant 1970: 48.

[8] For Maeterlinck poetry is 'a detour, and never speaks face to face [...]. It is the provisional mask behind which the faceless unknown fascinates us': McGuinness 2000: 96.

the body from the space, and their techniques therefore were strongly based in the suggestive potential of poetry, in the ear more than the eye. Strategies employed or suggested for this included a monotone, anti-expressive delivery, variously described by recourse to verbs such as 'psalmodize', 'intone', and so on; a focus in delivery on 'verbal orchestration' that emphasized the musical qualities of the text, especially vowel sounds, and the eradication of 'the theatrical voice' in favour of 'the poetic voice' (Deak 1993: 173, and cf. 54). Since the actor's body remained necessary, strategies for its depersonalization included the use of formal, hieratic gesture and slow movement with a 'ritualized' feel; the use of shadows rather than the fully corporeal human actor; the use of heavy make-up that removed the actor's identity and resembled a mask, and full-length costumes that hid the actor's physical particularities. And if the distraction of the actor's presence could not be overcome, then the complete disappearance of the human presence from the stage was recommended. Deak suggests that the techniques developed by Symbolist actors may have been influenced by the late nineteenth century puppet company, *Petit Théâtre de Marionettes*, and that, overall, Symbolist theatre 'raised the issue of representation in such a way as to put the existence of the live actor into question' (Deak 1993: 171, 174–5).

*Appia's Stairs – Platforms for the Body*

The translation of staircases from Symbolist painting to actual three dimensional stairs in theatrical space was principally the innovation of Adolphe Appia,[9] whose reforms, with those of E.G. Craig and others, helped transform the proto-Modernist theatre of the Symbolists into Expressionist and other 'Modernist' theatre forms. Whereas Wagner (the staging of whose operas Appia was initially so intimately associated with) through music, and the Symbolists, through poetry, had sought to achieve a spiritual renewal by eradicating the body from two-dimensional, illusionistic space, Appia used music to bring the body back into the space in a dynamic and plastic relationship with spatial depth and volume. This was Appia's principle gift to the twentieth century. A dynamic simplicity is evident in reproductions of

---

[9] Appia was no doubt drawing on the work of others. Several sources report that Appia was influenced by the use of platforms and multi-level floors at Anton Hiltl's Brunswick Court theatre: see Beacham 1987: 10. I have not been able to locate any further information on this.

Adolphe Appia's so-called 'rhythmic spaces', designs drawn from around 1909 while working with Jacques-Dalcroze at Hellerau (Bergman 1977: 328). In them you can see the prevalence of actual platforms and stairs, an enormous simplicity of conception, and the sense of monumentality and volume I mentioned earlier in connection with Wagner and the Symbolists. Dalcroze founded a method of learning and experiencing music through physical movement, and what was rhythmic about these spaces was that they were designed for this musically motivated movement, as well as to energize the physical space. By 1912 the studio at Hellerau had been adapted to the design of these drawings, with a single room divided into three sections of roughly equal length: steeply raked stairs and platforms, open floor, and audience seating.

Images of performances in this new studio reveal the dynamism of the actors' physical work on these stairs, and it is clear that in Appia's conception stairs and the bodies were intimately linked. In stark contrast to his early Symbolist influences, Appia considered the actor moving through space as the first cause of theatre, the primary element that all other elements of space and light must relate to in order for theatre to be 'living'. In *The Work of Living Art* he emphasizes that the weight and rigidity of inanimate forms in space are as essential to a living theatre as the human body. How different this is to the floating softness of Symbolist painting and bodiless poetry! As Appia writes, 'To receive its portion of life from the living body, space must oppose this body [...] opposition to the body gives life to the inanimate forms of space' (Appia 1960: 27). We can see this living relationship most clearly in the bodies on stairs of his rhythmic spaces, because of 'the obstacle they [the stairs] form to free walking, and the expression they give to the body' (Appia 1960: 25). The rigid step '[a]wait[s] the foot only to resist it, to throw it back at each new step, and to prepare it for a new resistance; through its rigidity, such a surface involves the whole organism in the spontaneity of walking. By opposing itself to life, the ground, like the pillar, can receive life from the body' (1960: 29). Elsewhere he writes about the platforms: 'a style suitable for establishing the value of the human body under the control of music' (1960: 112).[10]

[10] In 1906 Appia wrote to Dalcroze: 'the life of the body tends towards anarchy and therefore towards grossness. It is music which can liberate it by imposing its discipline upon it': Appia to Dalcroze, May 1906 (cf. Appia 1960).

The plasticity and dynamism of light, which for Appia is intimately linked to stairs, bodies, and music, is another of his most lasting legacies. Appia employed light as 'visual music' (Bergman 1977: 325), sculpting the plastic stage with it, unifying objects and colours, and emphasizing the dramatic values of the body. Appia differentiated 'diffused light' simply to allow vision and 'living light' that sculpted the living actor with its sharp and tight rays (Bergman 1977: 324–5). In an article he wrote in 1911, he writes of 'light without which there is no plasticity; light which fills the room with brightness and moving shadows […] And the bodies, basking in its animating atmosphere, will find themselves in it and greet the Music of Space' (cited in Bergman 1977: 329).

The shift that begins in Appia's work, in other words, is that the body is no longer ascending the celestial staircase into Symbolist exile, but placed centrally once more in a stage space dominated by stairs.

*Expressionist Stairs – the Body Descending*

Appia's work, along with that of Edward Gordon Craig, laid the foundations for the physical dynamism of German Expressionist theatre of the second and third decades of the twentieth centuries, in which stairs were used so often in theatre they became almost a *sine qua non* and, for some, a cliché. And it was in Expressionist theatre that the exploitation of the actor's dynamic physicality reached its peak in Modernist theatre, and directional lighting was most dramatically employed. Craig's resistance to the corporeality of the actor is initially marked, and like Maeterlinck and other Symbolists, he advocated the use of puppets to replace actors, because the actor did not possess the physical precision required by art. But Craig's principle legacy to Modernist theatrical space is his concept of moving screens that could provide dynamism to the stage like Appia's light and – after he fell in love with the dancer Isadora Duncan, perhaps – the moving body. Moving platforms were impractical, but screens could be moved to increase the architectonic liveness of the stage space. Some of Craig's drawings also reveal a preoccupation with stairs (see, for example, his 1905 Steps series, some of his 1907 'Scenes', and his 1909 designs for Macbeth). Indeed in a catalogue to an exhibition of some of these designs in England in 1912, Craig used the action of climbing steps to distinguish drama from literature, where the action was only described.[11] There was something about real steps that spoke to Craig of three-dimensional theatre. Stairs and platforms also made their mark in the Russian avant-garde stage. Alexander Tairov, at the Kamerny Theatre in Moscow, for example, expressed similar sentiments to Appia about the value of stairs and platforms.[12] In Expressionist theatre the staircase appeared from the very beginning. In Reinhard Sorge's *Der Bettler*, recognised as the first truly Expressionist play, the 'poet' descends the staircase at the end of the play to a deserted lower level.[13] There were many other uses of stairs in Expressionist theatre, for example in Alfred Roller's, 'Faust', part 2, at the Deutsches Theater, Berlin, in 1911, for Act III of Georg Kaiser's *Die Bürger von Calais*, written in 1913 and staged in 1917, and for Ernst Toller's *Masse Mensch*, directed by Jürgen Fehling at the Berlin Volksbühne in 1921 (Kuhns 1997: 211).

But stairs and steps in Expressionist theatre is most strongly associated with Leopold Jessner, who used them so often that they became known as the *Jessnertreppen*, or 'Jessner's Steps'. The stairs first appear in his work in 1920 in the Fourth Act of *Richard III*. Patterson's description is worth quoting:

> As the curtain rose at the end of the interval after the third act, a monumental flight of blood-red steps was revealed, its base filling almost the whole breadth of the stage, rising in three narrowing sets to just below the height of the wall [...]. Richard, in a long crimson robe, slowly mounted the red steps [for his coronation] through two lines of bowing henchmen. As he reached the top to ascend his throne, the red of the steps was linked to the red of the sky by the crimson of Richard's gown, as though an electric charge of evil had lept the gap between heaven and earth [...]. Later the same steps were used as the battlefield on which Richard assembled his warriors, on which Richard passed the nightmare-filled hours before battle, and on which the battle itself took place. (Patterson 1981: 93)

---

[11] The quote is from the catalogue of an exhibition of drawings and models for *Hamlet, Macbeth, The Vikings and Other Plays*, by Edward Gordon Craig, City of Manchester Art Gallery, November 1912, no.162, cited in White 2006: 97 and 103, note 10.

[12] See Tairov's comments on stairs and platforms at his Kamerny Theatre (van Baer 1992: 182).

[13] 'It is a powerful image of the artist turning from the exploitation of his art to descend wearily into the wilderness': Patterson 1981: 55.

After this production, Jessner was so taken by the steps that he just kept on using them, to the eventual chagrin of even his principle actor, Fritz Kortner.

Stairs were the tool by which Jessner wanted his actors to *embody* the idea of the production, not just represent or suggest it (as a Symbolist might do). The physical expressiveness of the actor's body is, apart from the stairs, the most characteristic aspect of the Expressionist acting event. Kuhns argues that 'the co-operative efforts of the productions' artists', including scenic and lighting designers, 'converged definitively on the body and voice of the actor [...] whose performance in turn infused the stage environment with great energy' (Kuhns 1997: 2–3). Drawings and photographs of productions such as Hasenclever's *Der Sohn* (1916), Sorge's *Der Bettler* (1917), Toller's *Masse Mensch* (1921) and others, reveal the dramatic use of directional lighting, the contrast between static and dynamic postures, contorted bodies, and emphasize how far the use of the actor's body here has shifted from the hieratic and ceremonial Symbolist theatre. And although ecstatic states were not characteristic of all kinds of Expressionist performance, they were strongly associated with the movement,[14] and formed a central focus of Expressionists' own theory. *Ecstasis* involved 'stepping outside of oneself [...] to become the embodiment of poetic form'. By removing unnecessary elements from the stage and the body, the actor sought to reveal and transform his soul, considered to be 'the only reality' (Gordon: 1975: 35). The *Jessnertreppen* were designed to assist in this great aim. Together with strongly directional lighting and dark shadows which sculpted the actor's ecstatic body, the stage space, so different to the ethereal and transparent Symbolist stage, was rhythmic and thick with resistance, something that was felt, as the dancer Mary Wigham put it, as if one were in water (Kuhns 1997: 92). Expressionist stage space did not reflect the external world but set about 'reconstituting it

[14] Many commentators differentiate between three kinds of Expressionist theatre: the *Geist* performance, the *Schrei,* 'scream' or 'ecstatic' performance, and the *Ich*, or 'I' performance, but Kuhns argues that ecstatic performance was present 'to some degree in every Expressionist approach to acting' (90). Mel Gordon differentiates between the aims of Expressionist playwriting and performance (Gordon 1975: 34), and Patterson (1981: 48–59) focuses his discussion of Expressionist theatre on two main tendencies, Abstractionism and Primitivism, in which the tendency towards an immediate, intense and ecstatic performance style, a primal scream in a more chaotic and distorted space, was more marked.

poetically from within' (Kuhns 1997: 92), and Gordon notes that the *Jessnertreppen*'s contribution to this aim was to signify 'the relationships between characters and their individual psychic states; it increased the actor's plastic possibilities, allowing him to be more easily perceived in depth; it rhythmically heightened the impact of slow, fast or disjointed movements; and it created a novel esthetic unity that was thought to be lacking in other Expressionist productions' (Gordon 1975: 50).

The point I want to stress here is that, just as they were for the Symbolists, the stairs represented and expressed a metaphysical position. Kuhns reports on an interview in which Jessner 'cautioned critics against regarding the staircase as merely a stylistic signature. Rather, it was simply the most effective setting for playing the "mythic events" which comprise the "idea" of a play' (Kuhns 1997: 210). As Jessner himself is reported to have said, 'the erection of the steps – as an autonomous architectural element – meant altering the base of the stage in accordance with its new function, which was now no longer to reproduce different rooms and landscapes but to be the abstract setting of mythical events' (Patterson 1981: 94). This was Jessner's *Motivtheater* (theatre of motifs), one in which the aim was to embody the central production concept on stage. According to Kortner, the stairs represented Jessner's 'Weltanschauung', and it is instructive that for Kortner, who believed the steps were originally his idea, not Jessner's, the stairs, and the action of climbing them, were an image of 'a career' in which one rose 'right up into the dizzying heights'. Patterson sees the steps as giving concrete form to 'the transcendent quality of Expressionism [...] a correlative of the soaring lyricism and philosophical search for a higher reality' (Patterson 1981: 94, cf. Kuhns 1997: 210, 197, 94). According to a contemporary German theatre critic, Alfred Polgar, the steps were infused with a Platonic metaphysic: 'They narrow towards the top and are free-standing in space. A sign that we are not to regard them as steps but as a vertical playing surface which we imagine stretching into infinity [...]. This is surely the Platonic idea behind Jessner's steps [...]. The performance gains a new dimension; the characterless movement to right or left is replaced by extremely meaningful moves up or down' (Patterson 1981: 94).

While I agree with the general tenor of these statements, the difference between the Symbolist and Expressionist use of steps is precisely one between a neo-Platonic and a post-Nietzschean metaphysic. The photographic reproductions of Jessner's steps make this amply clear. While Symbolist depictions of stairs lead ever upwards into an increasingly ethereal mist, the steps used in Jessner's *Richard III* lead to the top of a wall which doubles as a rampart, and behind this narrow level is another wall that quite clearly cannot be climbed. There is no virtual, mystical space beyond the top of *those* stairs. Central to Expressionist theory was the value of Man, and the search for personal spiritual renewal and oneness with 'the true, inner ecstatic reality of life' (Gordon 1975: 35). What drove the Expressionist actor was not a desire to ascend the celestial staircase out of this world, as in Symbolist theatre, but a 'longing to be fully and irresponsibly alive' (Kuhns 1997: 92). A regenerated soul meant that the actor (and hopefully individuals in the audience) would experience a renewed sense of oneness with their fellow men as he attempted, in Georg Kaiser's phrase, to become 'dissolved in humanity' (Kuhns 1997: 30–1). The regeneration of society that Expressionists as much as Symbolists sought was now to be achieved via the body in actual, material space.

The strong trace here of Schopenhauer's Will and Nietzsche's Dionysian life force is unmistakable. As Kuhns argues, from Schopenhauer Expressionist acting derived the notion of Will, that non-rational force 'which disguises itself in phenomenal experience and impels the universe onward with the force of its own ruthless self-assertion' (Kuhns 1997: 28–9). Schopenhauer had insisted that the whole body is nothing but objectified will, that the movement of the body and Will were one and the same thing (Kuhns 1997: 30). Nietzsche gave to Expressionist theatre his Dionysian life force, a more positive spin on the pain of being alive and a metaphysical – or rather anti-metaphysical – model for ecstatic performance connecting to a naturally regenerating essence, the *élan vital* of specifically communal life. Writing in Germany during the latter period of Expressionist theatre (his masterpiece, *Being and Time* was published in 1927), Martin Heidegger claimed that art was an 'intuition of Being', a means of knowing 'what it is to be', and for the Expressionists, as for Heidegger, 'being' for humanity was *being there*, spatially and temporally 'in the world'.

I have argued that the changing conception of 'what it is to be human' that is evident in Symbolist and Expressionist theatre is intimately connected to the way that painted stairways in Symbolist art became actual stairways and multi-level stage floors in the work of Adolphe Appia, the Russian avant-garde and German Expressionist theatre, and that this development involved the emergence (or re-emergence) on these stairs of the energised actor's body. In an illuminating article called 'Stairways of the Mind', architect Juhani Pallasmaa argues that:

> Like painting and poetry, architecture is engaged in articulating and expressing the human existential experience. The art of architecture creates spatial and material metaphors of our fundamental existential encounters [...] An architectural metaphor is a highly abstracted and condensed ensemble that fuses the multitude of human experiences into a single image.
> (Pallasmaa 2000: 7)

The stairway, with its symbolism of connection both to more rarified realities above and darker and less appealing realities below, is deeply rooted in mythology, and is a particularly resonant example of such architectural metaphors. Stairs are embedded in the archaic space of our consciousness, an *essence prototype* or *embodied concept*, in Lakoff and Johnson's terms (1999: 20). Cooper reports that, in the mythology of many cultures, stairs connote the 'passage from one plane to another or from one mode of being to another' and 'the ascending power of man's consciousness passing through all degrees of existence'. She further comments that stairs and ladders represent 'communication between heaven and earth with a two-way traffic of the ascent of man and the descent of the divinity', and 'access to reality, the Absolute, the Transcendent, going from the unreal to the real, from darkness to light, from death to immortality' (Cooper 1978: 94).[15] Baldon and Melchior, in a book devoted to stairs and stairways as an architectural feature, suggest, '[i]t is not without reason that the stairway is called a *flight*, for by it, foot over foot, earthbound man may rise to the height of birds' (Baldon and Melchior 1989: 13).

Pallasmaa's examination of the metaphysics of stairs supports my analysis in this paper. In addition to the door, he suggests, 'the stair is that element of architecture which is encountered most concretely and directly by the body' (Pallasmaa 2000: 9). With this in mind, it is clear that Symbolist theatre, with its desire to eradicate the body from

[15] See also the *Herder Symbol Dictionary* 1978: 113.

theatrical space, could not have used actual flights of stairs in their performances, for to do so would have worked directly against their aims. Symbolism was the inheritor of a fundamentally anti-theatrical, anti-body and neo-Platonic vector in Western philosophy; transcendent Symbolism took the epiphanic aspects of Plato's rationalism towards mysticism, gnosis and the noumenal, and sought regeneration through adherence to classical simplicity of line and a belief in what Nietzsche called Apollonian form. The Symbolists remained in Cartesian mind space, separated at an objective distance from the stage, and stuck in a nineteenth century concept of two-dimensional, illusionistic stage design which 'does not submit to any embodied immersion in space' (Wiles 2003: 7). A more phenomenal vector, nurtured by Nietzsche and later Heidegger, fed into (and out of) Expressionism and other 'modernist', anti-metaphysical expressions. Expressionist theatre, after Adolphe Appia's innovations with stage plasticity and depth, and a strong focus on the dynamic actor, brought the body back into theatrical space with a vengeance in an attempt to reconnect with their fellow human beings through an experience of Nietzsche's Dionysian delirium.

The use of a monolithic metaphorical image such as the staircase is typical, in fact, of the stage design of the period I have been examining. Aronson points out that '[m]odern stage design has been characterized by the presence of a strong metaphorical or presentational image or related series of images'. There was a sense of singular, 'organic' and 'monolithic' unity about these images, aptly described by Jean-Francois Lyotard's definition of modernism as 'a meta-discourse […] making an explicit appeal to some grand narrative'. Modern design, says Aronson, functioned by creating a 'meta-narrative that attempts to encompass the world within a unified image' (Aronson 2005: 13–14). The staircase was such an image, and tracking its use in Modernist theatre provides valuable insight into the crucial developments during this time in theatrical practice, metaphysical beliefs and the understanding of human consciousness. These developments, I suggest, had been set up by Kant's so-called Copernican Revolution, in which he suggested that we could not know 'things in themselves' (*das Ding an sich*) because our perceptions of the world were determined by the structures of our mind. Although Kant continued to believe that things in themselves existed, in hindsight the insight he brought to the study of consciousness, metaphysics and aesthetics helped to set off a series of revisions that have led to the proposition that '[t]he Mind is inherently embodied. Thought is largely unconscious. Abstract concepts are largely metaphorical' (Lakoff and Johnson: 1999: 3). The developments in Modernist theatre might be seen as a kind of 'play within a play' inside this broader shift.

*Postscript*

Gordon recounts an amusing story about the perils of the Jessnertreppen in performance: 'at the premiere of *Macbeth* [in 1922], Kortner, in a 'possessed' state of mind, lost his footing on the stairs and went sliding down the length of the platform' (Gordon 1975: 50). But the postscript to this reading of the relationship between Modernist theatre, consciousness and metaphysics has a rather more sombre side, for Schopenhauer's pessimism lurked at the bottom of those stairs. Lee Simonson's 1922 description of the wonderful effect of the *Jessnertreppen* in *Richard III* (1920) is prophetic:

> How immensely the movement of the second part was enhanced by the staircase when Richard appeared at the summit, when his men in red and Richmond's in white moved up and down it with all the symbolism of opposing forces, groups mounting towards its apex in imminent struggle. And what a contrast to all heightened movement as Richard descends slowly at the end in utter lassitude, to dream his last dream at its base.[16]

Richard's lassitude and his lost dream was also that of the Expressionists, once the initial fervour of the movement had, by the mid 1920s, lost its optimistic side. The hoped for regeneration of German society lead only to Hitler, who in the early 1930s squashed them under his heel. Kuhns suggests that Expressionist performance 'located its historical efficacy in the inspirational power over audiences that allegedly emanated from an ecstatically "possessed" stage'. But he further suggests that the highly energized Expressionist acting event, metaphorized and embodied by the *Jessnertreppen*, was not powerful enough to move social behaviour in the way that Hitler's oratory influenced the nation. The terrible irony of the *Jessnertreppen* was that placed centrally in the Nazi performance of power were huge rallies 'where phalanxes of enrapt automatons gazed up at their

---

[16] Lee Simonson, 'Down to the Cellar.' *Theatre Arts Monthly* VI (April 1922), cited in Gordon 1975: 50.

Führer, who stood at the apex of a monumental flight of stairs' (Kuhns 1997: 90, 223).

*References*

Appia, Adolphe. *The Work of Living Art: A Theory of the Theatre*. Transl. H.D. Albright. Coral Gables, Florida: University of Miami Press, 1960.

Aronson, Arnold. *Looking into the Abyss: Essays on Scenography*. Ann Arbor: University of Michigan Press, 2005.

Baars, Bernard. *In the Theatre of Consciousness: the Workspace of the Mind*. New York: Oxford University Press, 1997.

Baer, Nancy van. *Theatre in Revolution: Russian Avant-garde Stage Design 1913–1935*. London: Thames and Hudson, 1992.

Balakian, Anna. *The Symbolist Movement in the Literature of European Languages*. Budapest: Akademiaikiado, 1982.

Baldon, Cleo, and Ib Melchior. *Steps and Stairways*. New York: Rizzoli, 1989.

Baugh, Christopher. *Theatre, Performance and Technology*. New York: Palgrave MacMillan, 2005.

Beacham, Richard. *Adolphe Appia, Theatre Artist*. Cambridge, New York: Cambridge University Press, 1987.

Beacham, Richard. *Adolphe Appia: Essays, Scenarios, and Designs*. Transl. Walther R. Volbach; ed. and with notes and commentary by Richard C. Beacham Ann Arbor, Mich.: UMI Research Press, 1989.

Bell, Michael. 'The Metaphysics of Modernism.' *The Cambridge Companion to Modernism*. Michael Levenson (ed.). Cambridge: Cambridge University Press, 1999. 9–32.

Bergman, Gosta. *Lighting in the Theatre*. Stockholm: Almqvist & Wiksell International; Totowa, N.J.: Rowman and Littlefield, 1977.

Berman, Marshall. *All That Is Solid Melts Into Air: The Experience of Modernity*. New York: Penguin Books, 1988.

Blackmore, Susan. *A Very Short Introduction to Consciousness*. Oxford: Oxford University Press, 2005.

Bohrer, Frederick. *Orientalism and Visual Culture Imagining Mesopotamia in C19 Europe.* Cambridge: Cambridge University Press, 2003.

Chalmers, David J. 'Facing Up to the Problem of Consciousness.' *Journal of Consciousness Studies* 2/3 (1995): 200–19.

Childs, Peter. *Modernism.* London: Routledge, 1999.

Cooper, J.C. *An Illustrated Encyclopaedia of Traditional Symbols.* London: Thames and Hudson, 1978.

Craig, E.G. *On the Art of Theatre.* London: Heinemann, 1968.

Deak, Frantisek. *Symbolist Theater: the Formation of an Avant-Garde.* Baltimore: Johns Hopkins University Press, 1993.

Dennett, Daniel. *Consciousness Explained.* Harmondsworth: Penguin Books, 1993.

Gordon, Mel. 'German Expressionist Acting.' *TDR* 19/3 (Sept 1975): 34–50.

Grant, Damian. *Realism.* London: Methuen, 1970.

*Herder Symbol Dictionary*. Transl. Boris Mathews Wilmette, Illinois: Chiron Publications, 1978.

Jullian, Phillipe. *The Symbolists.* London: Phaidon, 1973.

Kearney, Richard. *Modern Movements in European Philosophy.* Manchester, New York: Manchester University Press, 1994.

Kuhns, David. *German Expressionist Theatre.* Cambridge: Cambridge University Press, 1997.

Lakoff, George, and Mark Johnson. *Philosophy in the Flesh.* New York: Basic Books, 1999.

McGuinness, Patrick. *Maurice Maeterlinck and the Making of Modern Theatre.* Oxford: Oxford University Press, 2000.

Morrison, Simon. *Russian Opera and the Symbolist Movement.* Berkeley: University of California Press, 2002.

Pallasmaa, Juhani. 'Stairways of the Mind.' *International Forum of Psychoanalysis* 9/1–2 (July 6, 2000): 7–18.

Patterson, Michael. *The Revolution in German Theatre 1900–1933.* Boston: Routledge and Kegan Paul, 1981.

Pincus-Witten, Robert. *Occult Symbolism in France: Joséphin Péladan and the Salons de la Rose-Croix.* New York: Garland Publishing, 1976.

States, Bert O. *Great Reckonings in Little Rooms.* Berkeley: University Press, 1985.

Szondi, Peter. *An Essay on the Tragic*. Transl. Paul Flemming Stanford, Calif.: Stanford University Press, 2002.

West, James D. *Russian Symbolism: A Study of Vyacheslav Ivanov and the Russian Symbolist Aesthetic.* London: Methuen, 1970.

White, Christine. 'Smart Laboratories – New Media.' Alison Oddey and Christine White, *The Potentials of Spaces: The Theory and Practice of Scenography and Performance.* Bristol: Intellect Books, 2006. 93–104.

Wiles, David. *A Short History of Western Performance Space.* Cambridge: Cambridge University Press, 2003.

# Bridge, Mirror, Labyrinth: Shaping the Intervals of Calvino's *Invisible Cities*

Kim Roberts

> It is no longer necessary to maintain the distinction between introspective knowledge [...] and objective knowledge. There is only one type of knowledge and it is always linked to an observer, an observer submerged in a system or its proximity. And this observer is structured exactly like what he observes. [...] There is no more separation between the subject, on the one hand, and the object, on the other [...]. This separation makes everything inexplicable and unreal. Instead, each term of the traditional subject-object dichotomy is itself spilt by something like a geographical divide [...] noise, disorder, and chaos on one side; complexity, arrangement, and distribution on the other. Nothing distinguishes me ontologically from a crystal, a plant, an animal, or the order of the world; we are drifting together toward the noise and the black depths of the universe [...]. Knowledge is at most the reversal of drifting, that strange conversion of times, always paid for by additional drift; but this is complexity itself, which was once called being. (Serres 1983: 83)

Italo Calvino's *Invisible Cities* is an artfully arranged constellation of urban visions, framed by the dialoguing ruminations of a merchant and an emperor. It presents as a cosmography that contains within and infers beyond itself maps and models of multiple and multiplying worlds. The interplay of the cities of the novel and the discourse that interleaves them is geometrically finite in its composition, but infinite in its conception. Thematic strands of cities are introduced; they build and decline, systematically overlapping the theme that is to supersede them. A woven form is fabricated via a mathematical formula that seeks a declension symmetrical to its inception. Here geometry is a fetish concealing absence. Like a fugue, an apparatus for multiplying and ordering voices, this armature facilitates invisible connections between the cities. The structure is exquisitely crafted, an abstract model of inevitability, and yet it is that which escapes his elegant net which haunts and fascinates Calvino – teeming complexity, flux, the abysmal void – that which is unspeakable, invisible.

This paper is concerned with relational space and spatial relations within Calvino's combinatory novel, *Invisible Cities*. It is an attempt to document the thematics of connectivity and separation operating upon the Calvinian *cogito* and the fluctuating cosmos in which it seeks to know, possess and locate itself. It is concerned with the spaces, or intervals, opened up by Calvino between ideas, objects and between his own literary authority and that of his readers.

The 'interval' is a pervasive spatial and conceptual phenomenon within Calvino's writing. It is the underlying topographical condition of all of his fiction and the substrate with which all secondary themes and spatial modes contend or attempt to address. The interval is the duration or relative span between one recognisable event, thing or phase and another. One may conceive of interval in terms of an interval of time, of space, or even of form – a place of transitory morphology, a zone for shape shifting. In the space of the interval boundaries come into being – they present and arrange themselves, or alternatively, their presence and arrangement becomes discernable. Things – objects, people and regions – their differences and similarities, become apparent. Surfaces and depths, proximity and separation are mutually revealed, measured and articulated. The interval is a space of relationships and dynamic inter-relationship – physical and formal, social and emotional, intellectual and cultural. It is a space of potential for communication and understanding. Across the interval desire drives the impulse for overcoming its span. Conversely, it is the space within which any lack of communication becomes apparent. It can rapidly transform into a gulf in which desire and fascination turn to horror and repulsion.

Within *Invisible Cities* the interval is apparent at many levels and scales. It operates between the inset and interlocking structures of the novel: the numerically devised armature by which the cities are relationally positioned; the italicized dialogue between Marco Polo and Kublai Khan; and the individual city units themselves. It facilitates or indicates resonances between its different parts. These are connections planted within the space of the imaginary, between what is written and what is not written. They are connections intended to take root within the minds of individual readers.

Of all his works, Calvino considered that it was in *Invisible Cities* that he 'managed to say the most' (Calvino 1996: 71). As he suggested in his lecture 'Exactitude', in *Invisible Cities*, the space or 'tension between geometric reality and the entanglements of human lives' was focused through the serial 'reflections, experiments and conjectures' upon the singular but infinitely complex and fertile symbol of the city (1996: 71). His self-confessed preoccupation with 'the conflict between the chaos of the world and man's obsession with making some sense of it' (Calvino in Weaver 1989: 30) was the generative source of the combinatory structures which give form to the novel. Through the rigours of these multi-faceted devices *Invisible Cities* offers its readers not the dry logical sequence one might expect from such a from such a controlled schema, but a ludic diffusion of authorial and narrative hierarchy. It presents a 'network in which one can follow multiple routes and draw multiple and ramified conclusions' (Weaver 1989: 30).

For Calvino, the value of literary precision, 'exactitude', has two impulses. The first of these is the schematization of events and experience into abstract models – an analytic, or quasi-scientific vision. The second is the intimate engagement of language with the world of things – a vision that is phenomenal and sensual. Calvino equates these two 'divergent paths' with two forms of knowledge, between which he must oscillate:

> One path goes into the mental space of bodiless rationality, where one may trace lines that converge, projections, abstract forms, and vectors of force. The other path goes through a space crammed with objects and attempts to create a verbal equivalent of that space by filling the page with words, involving a most careful, painstaking effort to adapt what is written to what is not written, of the sum of what is sayable and not sayable. (Weaver 1989: 75)

In *Invisible Cities* values, concepts and identities reveal themselves as multiple. The quest for a fixed and singular meaning is abandoned in favour of a set of highly differential narrative directives, each of which is irreducible to its own relatively incomplete and transitory textual existence. It is implied that the novel itself is, like anyone of the many cities which make up its parts, only a piece within a larger system, a larger universe, which, as Calvino wrote in *Mr Palomar*, is 'perhaps finite but countless, unstable within its borders which discloses other universes within itself' (Calvino 1985: 33).

This kind of utopian impulse is linked to the interval space which figures in the relationships Calvino portrays between the characters of his fiction and also between them and the cosmos in which they exist. If we consider – following a line of thought developed by the theorist of utopia Aurel Kolnai – that the key proposition of any utopian model is of life devoid of alienation, the connection between the interval and the almost compulsive production of discursive utopian models becomes apparent. The utopian temperament is highly sensitive to alienation, to the alterity that appears between the universe and the individual. The utopian program of non-alienation is aimed at a systemized smoothing over of the gap between the 'I' and the 'not-I'. This is generally achieved by way of a unifying Absolute, creating a second level super-alienation which perpetuates unthinkable violence on reality.

Calvino's characters in *Invisible Cities*, as in his other works, are quite distinctly marked by alienation. They are characterized, as Albert Sbragia put it, by 'an acute hypersensitivity to the disorder of the world' (Sbragia 1993: 272).[1] The substance, however, of the distinctly spatial disorder of alienation anxiety in this case extends beyond the 'parochial limits' (Kolnai 1995: 176) of the individual's relationship within the wider context of family and society addressed by most utopian models. Where most such models are focused upon the ideal relationship of the individual with the wider community Calvino's utopian impulse is not concerned with simply producing as ideal model for reality, or even an ideal self to dwell in it. It is concerned not with a static and unified 'utopia', but 'a utopian charge of energy' (Calvino 1989: 247). It is this charge, the restless movement from model to alternative model that allows a responsiveness and growth of ideals. For Calvino, ideological sensitivity and dexterity requires humanity to continually 'enlarge the sphere of what we can imagine' (1989: 247). New worlds must be constructed in the imagination, be thought out in all their details according to other values and other relationships (1989: 252) in order to perpetually generate new values, to create new stories.

Like many utopian projects, the written worlds of Italo Calvino are a response to the disorder of the universe. They are conceived as

---

[1] Sbragia extends this disorder to Calvino himself, comparing him with the figure of Mr. Palomar.

systematic articulations of spatial relations between discrete objects. These 'pieces' have the virtues of lightness, separability, ease of multiplication.[2] They lend themselves to infinite reordering and, as a result, the continual recreation of narrative meaning. Between the individual entities that combine to make up his universes Calvino draws out points of resemblance, pattern and bridges. He creates mutable lines of connection which paradoxically illustrate separation, the void and/or the teeming complexity they transverse.

In his 1967 essay 'Cybernetics and Ghosts', Calvino described his combinatory literature as a means to tap into 'an unexpected meaning or unforeseen effect which the conscious mind would have not arrived at deliberately: an unconscious meaning, in fact, or at least the premonition of an unconscious meaning' (Calvino *Cybernetics* 1989: 21–22). Combinatorial play, for Calvino, had the virtue of fracturing the authority traditionally granted to writers and offering instead a critical power to the reader. In the production of combinatorial literature, meaning – potentially slippery, if not in fact arbitrary – would only become substantiated in the reading. In order for such literature to become 'charged'[3] the reader must become a conduit between text and the 'hidden ghosts of the individual [reader] and of his society' (Calvino *Cybernetics* 1989: 22). In 'A Utopia of Fine Dust', written in 1973, Calvino directly connects the 'logico-fantastic machine' of combinatory literature with his utopian pursuits. Now 'the machine' produces not strict combinatorial output as much as critical model, after model, after model – seeking moral change:

> In a word utopia, not as a city that can be founded by us but that can found itself in us, builds itself brick by brick in our ability to imagine it, to think it out to the ultimate degree; a city that claims to inhabit us, not to be inhabited, thus making us possible inhabitants of a third city, different from utopia and different from all the habitable or uninhabitable cities of today; a city born of the mutual impact of new conditionings, both inner and outer.
> (Calvino, *On Fourier* 1989: 252)

---

[2] I am indebted here to Kathryn Hume's *Calvino's Fictions: Cogito and Cosmos* and her delineation of Calvino's handling of the material world in terms of minimal objects and flux.

[3] Calvino uses this term in discussing the revolutionary imaginary potential of both combinatorial literature and utopianism.

Combinatorial particles and the characteristically orderly cosmographic frameworks lend form to Calvino's later literature and are party to his project of proscribing universal order. Simultaneously, they are placed in tension by suggestions of a chaotic and much less concisely mappable universe. This is a universe that sweeps around and through these neat geometries, tugging and twisting them, folding them back upon themselves. This other[4] universe, 'countless, unclassifiable, in a state of flux' (Calvino *Cybernetics* 1989: 17), represents the failure of human ordering systems and the ability of the universe, in all its complexity, to elude them. Flux is largely a dystopian element in Calvino's cosmography. It, contesting Kolnai's theory, erases alienation, dissolving individual consciousness and identity. Aligning himself with the protagonists of his fiction, Calvino confesses his own vertigo in the face of chaos, 'a mess of vague and indeterminate lines', or 'a shapeless avalanche of events' (Calvino *Cybernetics* 1989: 17). Just as anxiety is present in his literature where a unit is irrevocably alienated from a system, an even greater anxiety is associated with the potential loss of individuated identity of any one unit, its consumption by a swirling 'sea of objects'. Mixed with this anxiety is fascination, desire.

Between the perceived disorder of the world and the rationalizing structures attempting to map and order it, exists a realm that holds a great, if largely unspoken power in Calvino's work. This realm is the gap, or interval which exists between them as the *outopic*, the displaced site of their difference. These concerns with the fission between chaos and cosmos partner several other obsessions, all of which are characterised by a similar space of difference. For Calvino there is a disjunction between the universe and its parts, between its parts and their ideation, and between alternative modes of ideation. Within these tensions are more specific relationships, likewise underscored by conceptual distance or absence. Foremost amongst these are the linguistic gap between language and phenomenal things within the experienced world, between fantasy and reality, and the division between the 'I' and the 'not-I' so fundamental to western

[4] These two universes, or visions of the universe are widely accepted to characterize Calvino's work, and are termed 'Cosmos' and 'Chaos' respectively by most commentators. For discussion of Chaos and Cosmos in the work of Calvino, see Hume, 'Grains of Sand' (1992: 72–85), Hume, *Calvino's Fictions* (1992) and Sbragia (1993: 283–306).

metaphysics. In his writing, the interval between the abstract model and the figurative model, between the constructed identity 'I' and its perceived universe is inherently spatial and inherently erotic. Further, it is characterized by a lack of place which has a potential that could be either utopian and generative or nihilistic and destructive (cf. Calvino, 'The Pen' 1989: 294).

Calvino is acutely conscious of the fact that any single model, being utopian and thereby of no-place, is necessarily redundant in advance of its creation. However, for Calvino no model ever exists solely in isolation but is understood as a system within a system, making any individual model never entirely disposable. It is within this dynamic space of shifting forms, changes of state and overturning ideals, that the power, challenge and beauty of Calvino's cosmographies lie. That is, in the mental and aesthetic no-places formed in the spaces between and the processes of the serial breakdown, re-propagation and re-positioning of their many models – the *ars combinatoria* which approaches an eternal return.

In *Invisible Cities* three themes or spatial modes are discernable which specifically consider the different ways in which the geographic divide or conceptual gap of the interval may be tackled. These modes are devices that investigate the nature of this space, whatever its form, and are posed as tentative and transitory solutions to the problems it poses.

The first of these is concerned with drawing lines or threads as direct links between ideas or objects. It also involves the construction of physical conceptual bridges by which to span the interval. The second mode broaches the interval through mirroring, by attempts to nullify or at least examine the distance through the mechanics of identification and differentiation, of semblance and distinction. In the third, the interval is grappled with by an intrepid tracing out of a journey – circular, spiralling or labyrinthine – through the pluralistic paths between banks, between the mutual, though not identical, imaginings of readers and writers.

As Calvino noted in his lecture 'Quickness', the entirety of Calvino's *oeuvre* is aimed at 'tracing the lightning flashes of the mental circuits that capture and link points distant from each other in space and time' (Calvino 1996: 48). Calvino seeks literary means bordering on the cartographic to draw lines, threads and bridges of connection that will measure the interval and make intelligible the interval – lines that will provide tentative solutions to the ontological vertigo or agoraphobic nausea produced by the abyss between the 'I' and the 'not-I'.

The bridges of *Invisible Cities*, however, do not simply or blindly cultivate the same toward the other merely to return to the same, that is, not without significantly problematizing this journey. Rather, they reveal their own tenuous and illusory nature as a 'ground of upsets', a story shattered, multiplied and rendered incomplete by a kaleidoscopic lens – bridges burnt in their telling.

In one of the most memorable moments within the italicized dialogue the following discussion takes place:

> Marco Polo describes a bridge, stone by stone.
> 'But which is the stone that supports the bridge?' Kublai Khan asks.
> 'The bridge is not supported by one stone or another', Marco answers, 'but by the line of the arch that they form'.
> Kublai Khan remains silent, reflecting. Then he adds: 'Why do you speak of the stones? It is only the arch that matters to me.'
> Polo answers: 'Without stones there is no arch.' (Calvino 1979: 66)

This passage is in many ways emblematic of the novel and the relationship between the numeric armature and the city units which it orders. In it the assemblage of stones implies the totalizing line which is the arch facilitating a conveyance between. However, in the discussion of the bridge between the merchant and emperor it is the structure – as either a series of components (the stones) or as a fully functioning whole (the arch) – that is in question. It is not difficult to see this figure as a metaphor for the structure of the novel itself, where the cities correspond to the stones and the novel's overriding form corresponds to the arch. This metaphor suggests models for two rhetorical schemas operating within the book: seriality (or metonymy), a cataloguing of (different) parts of the whole, and narrative continuity supported by metaphor, a process by which resemblances are drawn between the same.

The compulsive line or trajectory through the cities drawn by the armature emphasises a process of continual becoming rather than the self-satisfied reproduction of whole idea-objects. In his lecture 'Quickness', Calvino advises upon the necessity of narrative economy and tempo in the assemblage of events. It is an assemblage which he describes in terms of the 'puncti-form, connect[ion] by rectilinear segments, in a zig-zag pattern that suggests incessant motion'

(Calvino 1996: 35). In *Invisible Cities*, the city of Phyllis illustrates this narrative principal in spatial and experiential terms. In this city the traveller, like its inhabitants, 'follow[s] zigzag lines from one street to another', moving between the discernable points, visual and physical features of the city with which they tangibly engage:

> [...] a door here, a stairway there, a bench where you can put down your basket, a hole where your foot stumbles if you are not careful. All the rest of the city is invisible. Phyllis is a space in which routes are drawn between points suspended in the void: the shortest route to reach that certain merchant's tent, avoiding that certain creditor's window. Your footsteps follow not what is outside the eyes, but what is within, buried, erased. (Calvino 1979: 72–3)

In a similar, although less benign fashion, when the apple-faced boy of Calvino's early resistance fable 'The Crow Comes Last' raises his rifle to his shoulder, he wonders at the separation of things. He wonders that he finds himself so uncannily surrounded and 'separated from other things by yards of air' (Calvino 1983: 69). His gun is the talisman that for him orders this distance. When he takes aim, through the broad indeterminate field of the air 'a straight invisible line [is] drawn tight from the mouth of the rifle to the target' (1983: 69). When he shoots, the bewildering space between himself and objects that he considers already a part of him, inside him, projected on his retinal wall, is bridged: '[F]alse distances' are 'filled by a shot swallowing the air in between' (1983: 69). It is the dark lesson of this tale that the boy, seeking self-controlled connectivity and continuity between things, produces destruction of which he himself is oblivious and uncaring. The boy is like the agoraphobe who cannot step into open space without some form of guide to structure his path, be it a walking stick to tap before him or to follow the progress of another.[5] He zigzags through the landscape, shooting and moving in the wake of the bullet towards the fallen target. Upon reaching it he fixes on another target, shoots, and moves again. He avoids confrontation with the gaping generality of space by conclusively picking off one object at a time and moving in a straight, pre-determined line from point to

[5] See Paul Carter's discussion of agoraphobia as movement inhibition (Carter 2002). Accounts of Calvino's own self-confessed geographical neuroses and agoraphobic tendencies are to be found in his *Hermit in Paris* and 'Cybernetics and Ghosts'.

point. By eliminating the distance between himself and things in this way he eliminates or neutralizes their active difference.

Calvino systematizes the data of the experiential and imaginary world into minimal units, as represented by the cities. He then inserts them into a structure, as Kathryn Hume explains, as a way of 'stabilizing [...] experience of what is "out there". [...] If you can describe some small section of experience in a way that seems satisfactory, you have built a bridge between the I and the not-I, and by turning the not-I into words, you have remade it from your own mindstuff, turned into mental pictures, and further, turned it into a form that will provide others with mental pictures' (Hume, 'Grains of Sand' 1992: 81). Concise images such as those formed in the minimal units presented by the cites allow one to interiorize, consume and incorporate them into one's mental organizational systems. While no one model could hope to encompass the full extent or dimensionality of the universe, the swift passage through a number of finite images may succeed in producing a cinema graphic 'illusion of fluid movement and complexity' (Hume, 'Grains of Sand' 1992: 81).

In *Invisible Cities* the thematics of mirroring are most intriguingly located within the italicized dialogue which passes between Marco Polo and Kublai Khan. This dialogue involves an ongoing process of reflection upon modes of imagination and representation: themselves reflective functions. As emperor, Kublai Khan is the formal unifier of the many lands he has conquered and collected together to form his empire. Paradoxically, this self-made whole cannot be perceived except through the external viewpoint of another. The Khan, as creator and singular symbol of the realm is the figure most alienated from it.[6] Through the eyes, memories, emblems, gestures and finally, words of the foreigner, Marco Polo, the Khan's empire is enumerated, interpreted and reflected back to him. In a similar fashion, Marco Polo's place of origin, the cradle of his imagination, Venice, is also experienced at a remove, poignantly re-fabricated in each new city he describes.

---

[6] The Khan is not alone amongst Calvino's characters in this regard, but of all of them it is with that other king of Calvino's late story 'A King Listens', with whom he has the most resonance. The king of that story describes his alienation in this way: 'The palace, when you ascend the throne, at the very moment when it became your palace, became alien to you' (Calvino, 'A King Listens' 1993: 40).

The Khan is seeking the resolution of a self, whereas Marco Polo seeks to deconstruct one. Between the two men are passed objects, gestures, words, cities, like dowries. While both offer different ways, different scales and paradigms by which to read them, it is perhaps the momentum of the transversal between that undoes them and lends them new potential.

Kublai Khan and Marco Polo have different measures of the same world. Their respective subjective existences are born of different environments, differing sets of object and grammatical relations and so seek variant methods of comprehending and confirming these existences. Polo, the Venetian traveller, measures his worlds by trade, by exchange and by the cataloguing of its parts. The Khan, ruler of the Tatars, measures his by conquest, by material and intellectual possession, and by the systemic encapsulation of the whole. The Khan seeks unity, only to find the diversity of the world overflows his models. Like the boy of 'The Crow Comes Last', he seeks a possession of the objects of his world by which they will never draw close and yet never draw away. Marco Polo, on the other hand, searches for himself where he is not and in what he is not, in an equation not of lack but of difference.

One could simply emphasize the separateness of these two positions in conclusively oppositional terms. However, trade and possession, the mental and the physical, the parts and the whole all imply inherently interdependent and symbiotic relationships. Here, it is the desirous and mirror-like co-dependence of the two men which activates the interval space of their difference and identity formation.

As merchant and emperor, exchanger and possessor, cataloguer and unifier, each 'draws', that is, attracts and perpetuates through representation, the other's desire just as surely as Escher's famous drawing of two hands, in which, as Calvino writes: 'Every line presupposes a pen drawing it, and every pen presupposes a hand holding it' (Calvino, 'The Pen' 1989: 292). Just as the Khan is alienated, Polo is always a foreigner, textually and narratively: Arriving at each new city, the traveller finds again a past of his that he did not know he had: the foreignness of what you no longer are or no longer possess lies in wait for you in foreign, unpossessed places (Calvino 1979: 25).

The difference between the two men is capacity for motion. Kublai Kahn, as emperor, is a fixed symbol of the realm seeking for himself a static image of self, and whose steadfastly symbolic imagination allows for only probable and sanctioned associations. Marco Polo, however, is professional traveller and exchanger of imagistic symbols. As the seeker of improbable combinations, he is always moving, he has no place. To the Khan, who has employed him as 'mirror' to obtain a coherent reflection of his empire, he presents a notoriously restless surface. In their discourse he upsets the Khan's every attempt at permanently synthesising knowledge.

In the dialogue it is Kublai who possesses, or at least seeks, a more definitive subjectivity. In the externality of the merchant's accounts the Khan searches for the means by which he may construct for himself an ego. He seeks an imaginary unity which might see the coherence of the scattered parts of his empire – his symbolic body, and systemize an 'I' to supersede the omnipresent and diffuse appellations of 'We' or 'Us'. There is a sense that the figure of the Khan in his role as emperor has passed into the symbolic without first acquiring an imaginary and thus is driven to employ Marco Polo to access this realm. This quandary of being too much a creature of the symbolic, without access to the creative recourse of the imaginary, is perhaps illuminated in that the first use of the 'I' within the first fragment of direct dialogue of the novel comes from the Khan when he asks Marco: 'On the day when I know all the emblems, […] shall I be able to possess my empire at last?' To which the Venetian merchant replies: 'Sire, do not believe it. On that day you will be an emblem among emblems' (Calvino 1979: 21).

His role of storyteller allows Marco Polo to pass between the boundaries formed by the structural categories of Calvino's text. His voice persuasively functions in the italicized dialogue although his presence within the space of the dialogue is frequently questioned – is he real or a function of the Khan's imaginary? His voice is implicated in the narration of the cities but in a timeless, placeless realm outside that indicated by the italic dialogue. The narration of *Invisible Cities*, whether within the dialogue or the city descriptions, does not comply with expected conventions. The cities are not stories, they possess only the most rudimentary narrative groundings ('Leaving there and proceeding for three days towards the east […]' 'In vain, great hearted Kublai, shall I attempt […]' [Calvino 1979: 11, 13]). The relating voice of these moments speaks from a time and place without relation to the other narrative indicators of the novel, even if we assume that the voice that tells the cities is Polo's (cf. Breiner 1988: 560). If this point is accepted it is not far to the concession that the 'voice' of Marco Polo habitually shifts away from the first person singular, the 'I', to the third person singular, the theoretical traveller, expressed in the impersonal or passive form. The 'I' is in this way diffused, broken down and opened up. It is converted not to a vehicle of the universal *everyman* so much as to an *anyone*, mutable and full of potential. Within the italicized sections there is a similar lack of concern with the narrative positioning of voices within an identifiable time and place.

The voice that speaks from the regions attributed throughout the novel in general terms to Marco Polo, is one of dislocation and mobility, of pointed fictionality and imaginary. It is not, to follow Carol James' use of Maurice Blanchot's term, a proper narratorial voice (*voix narratrice*). As narrative subject, it cannot be relied upon to situate the recounted story, or stories[7] in time or place, or to exert some conscious control over its unfolding. The voice of Marco Polo is, in this sense, aligned with Blanchot's narrative voice (*voix narrative*) of 'radical exteriority'. The neutral voice speaks from a place where the work is silent; it has no place in the work, but neither does it look down upon it. The narrative voice places the fiction it narrates in a non-place, not in the idealistic sense of u-topia, but in the sense of residing in a fictional place, the *unheimlich*. This no-place is always outside and missing, and as it cannot be part of the work, but is nevertheless there, 'supplementarily': 'the 'it' [...] designates 'its' place both as where it will always be missing and which therefore would remain empty, but also as a surplus of place, a place always in excess: hypertopie' (James 1982: 152).

Marco Polo, with his disembodied eye and constantly shifting perspective, his combinatory tactics and his restless moving between viewpoints, is aligned with Calvino and his attmepts at achieving a co-extensive imaginary through the devices combinatory literature. The figure of Marco Polo, pre-linguistic, expressing himself 'only with gestures, leaps, cries of wonder and of horror, animal barkings or hootings, or with objects' (Calvino 1979: 20), as a form of mirror

---

[7] James notes that the even the device of embedded stories-within-a-story is destabilized here, 'where the structure implies two narrators, an outer and an inner, but in the narration itself the unfolding of the cities denies the possibility of containment' (James 1982: 152).

image, is incorporated by the Khan into his own interiority to the extent that by the second chapter, the two alter egos undertake their discussion, propose their 'questions and solutions' silently. Marco's 'answers and objections' assume their position within a self-sufficient 'discourse already proceeding on its own, in the Great Khan's head' (Calvino 1979: 24). Here the relationship fuses into a symbiotic möbius.

The two 'sides' of the discourse are an implied production of the same mind in a process of concentric reflection in which 'Marco Polo imagined answering (or Kublai Khan imagined his answer)', or, 'Kublai Khan interrupted him or imagined interrupting him, or Marco Polo imagined himself interrupted', 'so Marco Polo could explain or imagine explaining or be imagined explaining or succeed finally in explaining to himself [...]' (Calvino 1979: 24–5).

Which mind is truly hosting this debate is itself debatable, for although the text does indicate that it is the Khan's head which operates as forum, it is surely the imaginary realm associated with Polo which conjurers the scenario, that imagines, and therefore creates the interiority of the dialogue. Further, it is certainly Polo to whom Calvino gives more philosophical credence when in reply to the Khan, who asks him what the use of all his travelling is: 'to relive your past?' or 'to recover your future?' he responds with the words: 'Elsewhere is a negative mirror. The traveller recognises the little that is his, discovering the much he has not had and will never have' (Calvino 1979: 26). Marco searches for himself where he is not and in what he is not, in an equation not of lack, but of difference.[8]

---

[8] This deconstruction of self undertaken by Marco Polo reflects strategies of *différance* proposed by Jacques Derrida. In the essay of this name in *Margins of Philosophy* (1968) this term was introduced as a 'grammatological' illustration of Derrida's own conceptualization of difference. The substitution of 'e' with 'a' constitutes an inaudible difference perceived only in the graphic trace. This disruption of the supposed transparency of language – in which it is assumed to mirror thoughts and ideas without corruption, precisely by making visible a hidden possibility is also apparent in this discussion of the reproduction of apparently identical visual images from *Of Grammatology*: 'There is no longer a simple origin. For what is reflected is split *in itself* and not only as an addition to itself of its image. The reflection, the image, the double, splits what it doubles. The origin of the speculation becomes a difference. What can look at itself is one; and the law of the addition of the origin to its representation, of the thing to its image, is that one makes at least three' (Derrida 1984: 36).

In the cities of *Invisible Cities* circling, spiralling and labyrinthine paths and spatial gestures add to the vocabulary of form by which the novel and its parts are fabricated. The journeys they create plumb the subconscious depths and articulate the external surface contours of the space in-between which is the interval.

The oneness, stillness and non-directionality symbolized by the pure circle is not a sign that is prevalent in the work of Italo Calvino. Its wholeness, centeredness and harmonious continuity are out of keeping with his tendency towards multiple forms and ideas. Where circles do occur in his literature they are contextualized in terms of the relationships they form. Concentric, cyclic, combative and transformative, the circular relationships of *Invisible Cities* undermine the universalizing hermeneutic associated with the geometric figure, and yet, its refracted repetitions are ghosted by yearnings for just such an ideal order. The circle in Calvino's fiction is always seeking of or compulsively propelled into change, fuelled by the uncanny energy of the discursive charge between elements of un-reconcilable difference.

In the circle and the cycle, Calvino wrestles with the sanctity and apparent incorruptibility of form represented by the unbroken line which inscribes it, and the clear distinction between inside and outside it signifies. Calvino rejects the circle as a formal emblem. Like Nietzsche's third and obliquely articulated eternal return of *Thus Spake Zarathustra*, Calvino's cities of combinatory returns are of the 'Different, the Dissimilar' (Deleuze 2001: 299), the uncanny utopian charge or Dionysian 'force' that creates models and, finding them wanting, goes on creating more. For, as Deleuze writes of the eternal return of Nietzsche:

> '[O]ne' repeats eternally, but 'one' now refers to the world of impersonal individualities and pre-individual singularities. The eternal return is not the effect of the Identical upon a world become similar, it is not an external order imposed upon the chaos of the world; on the contrary, the eternal return is the internal identity of the world and of chaos, the Chaosmos. (Deleuze 2001: 299)

In Calvino's fiction the circle always desires other forms. That is, it desires non-fixity of form, other dimensions in spatialization and temporalization and the elimination of fixity in time and space. Therefore, in *Invisible Cities* we witness the flattening out of time and space into a single line of 'anti-narrative' in which forms are evoked but partially, inconclusively, where cities are dislocated geographically, and where narrative time is stripped of relativity.[9] As Calvino often finds, this is not a line in a single dimension describing a regular Platonic form, but one which sees the cycle passing through ever variant moments of space and time. It is the circle split from itself, peeled from the page – the spiral, which to the traveller, lacking a transcendental knowledge of its form, becomes a labyrinth.

Through remembrances, forgettings and conflations of the past and future the invisible cities circulate. Commencing with the grouping, 'cities and memory', and finishing with 'hidden cities', the categories are equally multifarious and elusive. Woven by the dancing line of a meander, they expand, contract and locate themselves in each other. In a city of memory such as Zaira, history is found to be contained – hidden, yet fully visible in the seamless surface and structuring of its physical fabric. The city, like the many other cities of the novel, provides a metaphor, at least part of one, for conceptualizing human knowledge and cognition. By extension it provides a model for the construction of consciousness, of the *cogito*.

Memory has no fixity. It threatens to mutate or slip away and yet it is found in the 'relationships between the measurements of its space and the events of its past [...] like the lines of a hand, written in the corners of the streets, [...] every segment marked in turn with scratches indentations, scrolls' (Calvino 1979: 13). The city Zaira, suffused with waves of memories, 'soaks [them] up like a sponge and expands' (1979: 13). The events of its past have measured the width and breadth of the city's expanses, have constructed and made sensible its edifices, its hidden places, and its architectural detail. Now, however, the events can be connected to their physical traces only in the tale told and re-told by the three old men, mending fishing nets on the dock. The physical markings and relationships – themselves alternative memories of the historical happenings – are reproduced ('for the hundredth time') in yet another form: oral narrative. Further, they are yet again 'associatively displaced'[10] in the

[9] Temporal disjunctions occur within the italicized dialogue, between the dialogue and the cities and between the cities themselves.

[10] Cf. Freud 1966: 52: '[I]nstead of the memory which would have been justified by the original event, another memory is produced which has been to some degree associatively displaced from the former one.'

account of Marco Polo (retrospectively addressed to Kublai Khan, it can be assumed), on page thirteen of Italo Calvino's *Invisible Cities.*

The cities, as evidenced in Zaria, are complex constructs of repressions (clandestine relations, abandoned illegitimate children, and the execution of the usurper) and replacements (physical traces of events, the return of the abandoned child as usurper, and the telling of tales). That is, they are recollections swathed in artifice. The 'repression' of Marco Polo's remembrances of Venice and their 'replacement' by memories of other cities are valuable because they represent 'in the memory impressions and thoughts of a later date whose subject-matter is connected with its own by symbolic or similar links' (Freud 1966: 66). *Invisible Cities* is characterized by this type of play of illusion – of evocative, tangible moments of presence amid absence.

While the 'replacements' of Calvino's combinatory and substitutional literary tactics (corresponding with the Freudian model and the example of his character, Marco Polo) appear obvious, the substance of personal 'repressions' is more difficult to trace. In consideration of Calvino's understanding of the transformation, distortion and erasure of the past by its telling they are considerably less important than the patterns of substitution which overtake them.[11] Calvino locates certain spaces of repression, in the gaps usually smoothed over in traditional literature, such as those between writers and readers. As noted by Marco Polo: '"I speak and I speak", Marco says, "but the listener retains only the words he is expecting [...] It is not the voice that commands the story: it is the ear"' (Calvino 1979: 106). This is also borne out in the gaps between memory and experience, experience and literature, as Calvino writes in the preface to *The Path to the Spiders' Nests*:

> Memory, or rather experience – which is the memory of the event plus the wound it has inflicted on you, plus the change which it has wrought in you and which has made you different – experience is the basic nutrition also for a

[11] Just as Marco Polo speaks of the erasure of Venice that he fears may occur or already has occurred in speaking of it, directly or through speaking of other cities, Calvino in his 1964 preface to his first novel, *The Path to the Spiders' Nests* (1998: 28) wrote of the capacity of literature to erase memory: 'The first book instantly becomes a barrier between you and that experience, it severs the links that bind you to those facts, destroys your precious hoard of memories [...] Your memory will never again recover from this violence that you have done to it by writing your book.'

> work of literature (but not only for that), the true source of wealth for every writer (but not only for the writer), and yet the minute it gives shape to a work of literature it withers and dies. The writer, after writing, finds that he is the poorest of men. (1998: 29)

This impoverishment, or emptying out of the author, makes of him a place dislodged from the fabric of the text, a non-place – utopian, yet also uncanny. As Calvino writes in 'I Also Try to Tell My Tale', *The Castle of Crossed Destinies* (evoking the 'doctrine of dreams' of Sigmund Freud and Roland Barthes's 'death of the author'): 'In writing, what speaks is what is repressed' (Calvino, 'I also Try' 1979: 98). That which is repressed in the works of Calvino – the voice of the author – however, is not conclusively erased. Instead 'an author's shade' is called up like Dante's Virgil, to accompany the reader's 'distrustful steps' (Calvino, 'I also Try' 1979: 100), to zigzag between discursive boundaries like a ghost or angel – a guide within the shadowy realm of their individual and shared imaginary.

*Invisible Cities* is a crucible of playful experimentation in which words and models are 'linked up to test their reciprocal fertility' (Calvino 1970: 93). Looping networks of relationships and interlocking structures form a labyrinth. Combinatory assemblages of literature and literary subjectivity are revealed as serial processes of orientation, disorientation and reorientation. Boundaries are called into question, narrative and authorial voices are fragmented and redeployed against each other. Cities are celebrated not merely as objects in and of themselves but for the magnetic charges of connectivity and separation that order their seriality.

In *Invisible Cities* labyrinthine seeking is revealed as a utopian vocation. For Calvino the solution to the puzzle of the labyrinth is not simply found in a flight from its depths, Icarus-like, towards a Platonic or Cartesian enlightenment, but in the continued exploration of its contours. While acknowledging the aesthetic pleasures of construing the world in terms of crystalline abstract models and mappings, he simultaneously embraces the irony of the inevitable limitations that compel the search for new representations.

Labyrinthine in its conception and execution, each region, relationship and object of *Invisible Cities* is enveloped by, or intertwined with, others. However, each in some way exceeds, each resists or denies the containment implied by such boundaries. The complex utopian vision presented by the novel is of a community of dislocated identities connected, tensioned and focused by a number of structural devices and allusions. The continuing oscillation between multiple forms of knowledge which these forms promote prevents stasis and produces a discursive utopianism of hybridity and co-extension. Wherever co-extension of one object, person or region into another occurs questions are generated and a re-evaluation is prompted. This chiasmus, crossing-over of boundaries entails a going-under, hidden challenges are revealed beneath the surface of a network of potential utopian models and potential selves in a veritable remodelling of the interval.

Cities are networks of connectivity, of combinatory association, of desire. Simultaneously, they are webs of discontinuity, of alienation and repulsion. They are the ever-changing yet eternal repositories of the traces and absences that form relationships and are formed by them. They see the accrual and remembrance of markings and movements, their lack, non-appearance, disappearance and forgetting. As Manfredo Tafuri suggests, the city is 'a place wherein the loss of identity is made an institution' (Tafuri 1990: 291).

As much as any of its parts presents as a model of the city, *Invisible Cities* itself is such a model. With its shifting kaleidoscopic arrangement, with its movement through a series of visions and spaces enveloped and intertwined with other visions and other spaces it maps a state of continual chiasmus. It portends the possibility of a mutual space, a continually moving and re-informing moment and the cross-roads of the interval of the in-between.

## *References*

Breiner, Laurence. 'Italo Calvino: The Place of the Emperor in *Invisible Cities*.' *Modern Fiction Studies* 34/4 Winter (1988): 559–573.

Carter, Paul. *Repressed Spaces: The Poetics of Agoraphobia*. London: Reakion Books, 2002.

Calvino, Italo. 'A King Listens.' *Under the Jaguar Sun*. Transl. William Weaver London: Vintage, 1993. 31–64.

—. 'Cybernetics and Ghosts.' *The Literature Machine*. London: Picador, 1989. 3–27.

—. *Hermit in Paris*. Transl. Martin McLauglin. London: Jonathan Cape, 2003.

—. 'I Also Try to Tell My Tale.' *The Castle of Crossed Destinies*. Transl. William Weaver. London: Picador, 1979. 95–107.
—. *Invisible Cities*. Transl. William Weaver. London: Picador, 1979.
—. *Mr Palomar*. Transl. William Weaver. London: Secker & Warburg, 1985.
—. 'Notes Towards a Definition of the Narrative Form as a Combinative Process.' Transl. Bruce Merry. *Twentieth-Century Studies* 3 (1970): 93–101.
—. 'On Fourier.' *The Literature Machine*. London: Picador, 1989. 245–55.
—. 'Preface.' *The Path to the Spiders' Nests*. Transl. Archibald Colquhoun. Revised Martin McLaughlin. London: Jonathon Cape, 1998: 7–30.
—. *Six Memos for the Next Millennium*. Transl. Patrick Creagh, London: Vintage, 1996.
—. 'The Crow Comes Last.' *Adam, One Afternoon*. Transl. Archibald Colquhoun & Peggy Wright. London: Secker & Warburg, 1983. 68–73.
—. 'The Pen in the First Person.' *The Literature Machine*. London: Picador, 1989. 291–9.
Deleuze, Gilles. *Difference and Repetition*. Transl. Paul Patton. New York and London: Continuum, 2001.
Derrida, Jacques. *Of Grammatology*. Transl. Gayatri Chakravorty Spivak. Baltimore and London: The Johns Hopkins University Press, 1984.
Freud, Sigmund. 'Childhood Memories and Screen Memories.' *Psychopathology of Everyday Life*. Transl. James Strachey. London: Ernest Benn, 1966. 43–52.
Hume, Kathryn. *Calvino's Fictions: Cogito and Cosmos*. Oxford: Clarendon Press, 1992.
—. 'Grains of Sand in a Sea of Chaos.' *Modern Language Review* 87/1 (1992): 72–85.
James, Carol P. 'Seriality and Narrativity in Calvino's *Le città invisibili*.' *Modern Language Notes* 97/1 (1982): 144–161.
Kolnai, Aurel. 'Utopia and Alienation.' *The Utopian Mind and Other Papers: A Critical Study in Moral and Political Philosophy*. Ed. Francis Dunlop. London: Athlone, 1995. 176–182.
Sbragia, Albert. 'Italo Calvino's Ordering of Chaos.' *Modern Fiction Studies* Summer (1993): 283–306.

Serres, Michel. *Hermes: Literature, Science, Philosophy*. Baltimore and London: The John Hopkins University Press, 1983.

Tafuri, Manfredo. *The Sphere and the Labyrinth: Avant-Gardes and Architecture from Piranesi to the 1970s*. Transl. Pellegrino d'Acierno and Robert Connolly. Cambridge, Mass. and London: The MIT Press, 1990.

Weaver, William. 'Calvino: an Interview and its Story.' *Calvino Revisited*. Ed. Franco Ricci. Toronto: Dovehouse Editions, 1989. 17–31.

# Cartographers of Consciousness: Imagined Library Spaces in the Work of Haruki Murakami, Umberto Eco and Elias Canetti

Ewen Jarvis

In a photograph taken in 1984 by Magnum photographer Ferdinando Scianna (Scianna 1998: 16), the writer Jorge Luis Borges is seen seated in what appears to be the dining room of a hotel. A terrace upon which wrought iron garden chairs have been arranged is reflected in the plate glass window through which the photograph was taken, so that there is initially some confusion as to whether the subject of the photograph is inside or outside. His dark, double-breasted suit hangs loosely on his aging limbs, his hands are wrapped around a walking stick as thick as Moses's staff, and with a far away expression he appears to be staring through the veil of his blindness into the vast and fertile universe of his own imagination, and it is easy to suppose that what he sees there is a library.

However, being a man who spent much of his long life in libraries, it is likely that the realm into which he was gazing on that day in 1984 contained not one but many libraries. It may well have contained the Argentine National Library of which he was the Director; it may have contained the infinite library he describes in 'The Library of Babel'; and it may also have contained countless other libraries, both actual and imagined. And these libraries of memory and imagination would certainly have overlapped in places creating the sort of superimposition that one sees in Scianna's photograph, in which the interior space of foyer, the exterior space of the terrace and the unseen yet multidimensional space of the author's imagination can be said to coexist. What interests me here is the degree to which such movements and minglings of interior library spaces – which represent systems of ideas – can be said to mirror the workings of the human psychic apparatus which governs the ebbs and flows and composition of consciousness itself. Can internalized library spaces be seen as maps that can help us to explore the way we apprehend the world? In answering this question I have found it profitable to turn to libraries as they appear in literature.

I will be examining the imagined libraries of three modern novelists and considering what aspects of library space they have chosen to emphasise in their work. Umberto Eco's *The Name of the Rose*, Elias Canetti's *Auto-da-Fé* and Haruki Murakami's *Kafka on the Shore* all contain imagined library spaces, which exist to be entered and explored. In examining these imagined libraries, I will consider how they have been constructed, what the library spaces mean to the characters who inhabit them, what aspects of library space have been emphasized, and to what degree these imagined libraries allow us to apprehend the shape of the human mind and the nature of consciousness itself.

### *'The Name of the Rose': the Library as a Space of Secrets, Power and Paranoia*

The first library space I wish to enter and explore appears in Umberto Eco's *The Name of the Rose*. The narrated events take place in 1327. A young Benedictine novice, Adso of Melk, and a learned Franciscan brother, William of Baskerville, enter an abbey in the mountains of Northern Italy: an abbey in which the greatest library of Christendom is preserved. The library, which is housed on the top floor of a fortress like structure referred to as the Aedificium, is described to us in the following way by the novice and narrator Adso of Melk:

> While we toiled up the steep path that wound around the mountain, I saw the abbey. I was amazed, not by the walls that girded it on every side [...] but by the bulk of what I later learned was the Aedificium. This was an octagonal construction that from a distance seemed a tetragon (a perfect form, which expresses the sturdiness and impregnability of the City of God).
> (Eco 1983: 21)

Already, the physical form of the library can be said to reflect its nature. It is a library one must toil to reach, and its very shape speaks of impenetrability. Adso goes on to say that 'the southern sides stood on the plateau of the abbey, while the northern ones seemed to grow from the steep side of the mountain, a sheer drop, to which they were bound'. And he observes that from below the Aedificium seemed to 'reach up towards the heavens'. (Eco 1983: 22) This description seems to suggest that the library's impregnability is inextricably linked to vague promises of spiritual ascent and descent. And indeed later in the text the library is described as being 'at once the celestial Jerusalem and an underground world on the border between terra incognita and Hades'(Eco 1983: 184), and this emphasis on verticality is present throughout the text.

The threat of descent, in particular, quickly makes itself apparent, for hardly have the travellers arrived than they are informed that some days past a young monk was found dead 'at the bottom of the cliff below the Aedificium' (Eco 1983: 31), and at the request of the Abbot, the protagonists seek to discover the author of the crime, with the stern injunction that they are forbidden from entering the library itself. For they are informed that:

> The Library was laid out on a plan which has remained obscure to all over the centuries, and which none of the monks is called upon to know. Only the librarian has received the secret, from the librarian who preceded him [...] And the secret seals the lips of both men. (Eco 1983: 37)

In this way, the library is presented not only as an impenetrable space where only those deemed worthy have access to the 'divine word', it is also presented as a place of secrets. A librarian and his assistant are the gate keepers of the library and all it contains. As William of Baskerville observes: 'The place of forbidden knowledge is guarded by many and most cunning devices. Knowledge is used to conceal, rather than to enlighten' (Eco 1983: 176).

In *Crowds and Power*, Elias Canetti states that 'Secrecy lies at the very core of power' (Canetti 1984: 290). And he maintains that

> Secrecy retains its dual character [of at once protecting against and aiding in the perpetration of evil] in all the higher manifestations of power. It is only a step from the primitive medicine-man to the paranoiac, and from both of them to the despot of history. (Canetti 1984:292)

And since Eco's library is found to be ruled by Jorge, a blind clandestine despot, it is not surprising to find that Jorge suffers from the major symptoms of paranoia: delusions of persecution, unwarranted jealousy, exaggerated self-importance all worked into an organised system. So it is hardly surprising that Jorge also conforms to Canetti's description of a despot

> [in whom] secrecy is primarily active. He [the despot] is thoroughly familiar with it and knows how to assess its value and use it on any given occasion. When he lies in wait he knows what he is watching for and knows, too, which of his creatures he can use to help him. He has many secrets, for he has many desires; and he organises these secrets so that they guard one another. He reveals one thing to one man and another to a second, and sees to it that they have no chance of combining them.

'Power', Canetti holds, 'is impenetrable' (Canetti 1984: 292).

Jacques Derrida in *Archive Fever* observes that 'There is no political power without control of the archive' (Derrida 1996: 4). And indeed lust for power is central to attempts to control the library and to contain its secrets. At the end of his section on secrecy Canetti states that:

> in the long run, all secrets which are confined to one faction, or, still more, to one man must bring disaster, not only to the possessor, [...] but also to all they concern [...] Every secret is explosive, expanding with its own inner heat. (Canetti 1984: 295-6)

And Eco's invented library harbours a dangerous concentration of secrets.

First and foremost, there is the secret that seals the lips of the librarian and his assistant, which concerns the layout of the library; for as well as being aligned with the four cardinal points, the shape of the library also corresponds to the then known regions of the earth. Each room in its labyrinth of rooms is given a letter, and linked up these letters spell out the names of the then known portions of the globe: Anglia, Germani, Gallia, Hibernia, Roma, Yspania, Leones, Aegyptus, Iudaea, Fons Adae and finally in the secret centre the finis Africae. The library is in fact a mirror of the world.

Many allusive possibilities stem from this correspondence, but most importantly it lends legitimacy to a counter narrative which involves the book of Revelation and the end of the world; and this counter narrative is used by Jorge as a kind of mask or disguise beneath which his explosive secrets can be concealed. Canetti describes such a mask or disguise as follows: 'Hiding, or taking on the colour of its surroundings and betraying itself by no movement, the lurking creature disappears entirely, covering itself with secrecy as with a second skin' (Canetti 1984: 290).

In Eco's narrative, only the learned William of Baskerville is able to penetrate this second skin, and when it is eventually discovered that Jorge has been ruling the abbey and the library illegitimately from behind a veil of appearances; that he has been privy to the nature of the murders which have plagued the abbey; that he is responsible for the repression of the book which has caused so many deaths; only then is the scene ripe for the explosion of heat which ultimately destroys the library and the rest of the abbey with it. As Adso observes, 'the library had been doomed by its own impenetrability' (Eco 1983: 489).

For Jorge the library is a place to be partitioned and controlled. And when he comes across an element that does not 'articulate the unity of an ideal configuration'(Derrida 1996: 3) the urge to partition and control becomes what Freud has called among other things 'an instinct of destruction'. In speaking of this instinct for destruction Derrida, in *Archive Fever*, states that

> this drive which [...] [Freud] names here sometimes death drive, sometimes aggression drive, sometimes destruction drive, as if these three words were in this case synonyms. [...] this three named drive is mute [...] It is at work, but since it always operates in silence, it never leaves any archives of its own. It destroys in advance its own archive, as if that were in truth the very motivation of its most proper movement' (Derrida 1996: 10).

This death/aggression/destruction drive is what has overtaken the despotic Jorge. He, as the guardian of the documents, whom Adso refers to as 'the library's memory' (Eco 1983: 130) and whom Derrida would refer to as an *archon* (Derrida 1996: 2), in trying to maintain an order and a unity, ultimately brings about the destruction of the library. Derrida pairs the power of these archons, these guardians of documents, with what he calls

> the power of consignation which aims to coordinate a single corpus, in a system or a synchrony in which all the elements articulate the unity of an ideal configuration. In an archive, there should not be an absolute dissociation, any heterogeneity or *secret* which could separate (*secernere*), or partition, in an absolute manner. (Derrida 1996: 3)

However, in Eco's library the most troublesome secret is the existence of Aristotle's treatise on comedy, which does threaten to separate and partition in an absolute manner. And as such, when this secret stands in danger of being discovered, the three-headed-beast, the death/aggression/destruction drive, emerges to consume the archive.

### *'Auto-da-Fé': the Library as a Space of Pure Intellectual Activity and a Site of Contamination*

Another imaginary library which is the site of power plays of a different kind appears in Elias Canetti's *Auto-da-Fé*. Unlike Eco's monastic library, Canetti's imagined library is a private one, and its owner, Peter Kien, is a famous sinologist devoted to intellectual pursuits and the preservation of his formidable collection of books. His library is in his home and it is made up of four rooms of equal size. Natural light shines through skylights which he has had installed, but otherwise life beyond the library has been completely blotted out. In Peter Kien's lofty Apollonian world of ideas there is no place for any extraneous contact with outsiders.

In traversing this particular library space the words of London based artist and writer David Batchelor come to mind: words he used to describe the bleached and bare living space of an Anglo-American art collector:

> Inside this house was a whole world, a very particular kind of world, a very clean, clear and orderly universe. [...] It was a world that didn't readily admit the existence of other worlds. Or it did so grudgingly and resentfully, and absolutely without compassion. (Batchelor 2000: 10)

For Peter Kien the library is a scene of pure intellectual activity and any infringement by the outside world is looked upon as a contamination. His library, we are told, 'was situated on the fourth and topmost floor of No. 24 Ehrlich Strasse and the door of the flat was secured by three highly complicated locks' (Canetti 1984: 23). Canetti goes on to describe a space in which anything extraneous is looked upon with revulsion, a space in which 'no single superfluous article of furniture, no single superfluous person could lure him from his serious thoughts' (Canetti 1984: 23). It is a space in which even the windows have been walled up in order to avoid the 'temptation to watch what went on in the street' (Canetti 1984: 23).

So when Peter Kien's brutish housekeeper, Therese Krumbholz, begins to suspect that her master is concealing weaknesses behind his severe exterior, she contrives a way to improve her standing in the world at his expense. Their subsequent marriage becomes a war in which Kien slowly sees himself dispossessed of his library one room at a time, until he is eventually locked out of his apartment and forced to fend for himself on the streets; and ill equipped for such a life, he is taken advantage of by all and sundry.

A spatially interesting phenomenon takes place after he has been separated from his library; he begins to internalise a new library by utilizing his remarkable memory so as to carry an 'entire new library in his head' (Canetti 1984: 169). And thus, roaming the streets with the burden of the innumerable volumes he imagines are stored inside him, Peter Kien's never very strong hold on reality grows ever weaker. No matter how hard Kien tries to maintain, if only mentally, a space where his intellect can be free, the forces of greed, pride, self-righteousness – all the baser manifestations of corporeal existence – press ever harder upon him to contaminate his ideal.

In this bleak modernist masterpiece, the library is at once a home, a space set aside for pure intellectual activity, a site of contamination, a commodity, a symbol of power, and ultimately an inferno. Like the abbey library in *The Name of the Rose*, Peter Kien's library is destroyed in a conflagration, and these two conflagrations are linked to the madness of the patriarchs who have imagined themselves the library's owners; Jorge ends his life by eating the poisoned pages of the book he has tried to conceal, and Kien burns his own library so that no one else can take possession of it. And although they come to their end in different ways, their ends are similar as they both cease to be able to distinguish between reality and imagination and are consumed by an unattainable unity and purity that they ultimately fail to realize.

As Salman Rushdie has observed, '[i]n *Auto-da-Fé* no one is spared. Professor and furniture salesman, doctor, housekeeper, and thief all get it in the neck. The remorseless quality of the comedy builds one of the most terrifying literary worlds of the century'.[1] Canetti does succeed in making a hell of his invented library.

---

[1] This quotation attributed to Salman Rushdie appears on the back cover of the edition of *Auto-da-Fé* to which I have been referring.

### *'Kafka on the Shore': the Library as a Space of Liminality and Transformation*

The last invented library I wish to elaborate on is a library somewhat different from the European libraries of Eco and Canetti. The Japanese author Haruki Murakami often uses library spaces in his work, but his most developed library appears in *Kafka on the Shore*, a novel about a fifteen-year-old boy who runs away from his home in Tokyo and finds himself living in a small regional library. What differentiates Murakami's library from the two already described is that it is open and malleable, whereas the others are closed and ossified.

The protagonist's description of his journey towards and into the library is similar to Adso of Melk's description in that they both lend themselves to symbolic interpretation. After running away from home and booking himself into an hotel, Murakami's fifteen year old protagonist Kafka Tamura decides to visit the Komura Memorial Library, which he has seen in a magazine article and which the reader senses he has been drawn to. From the station to the library small signs 'pointing towards the library line the road' (Murakami 2005: 36), and when he arrives at the gates of the library he states that:

> Right in front of the Komura Memorial Library's imposing front gate stand two neatly trimmed plum trees. Inside the gate a gravel path winds past other beautifully manicured bushes and trees [...] with not a fallen leaf in sight. A couple of stone lanterns peek out between the trees, as does a small pond. Finally, I get to the intricately designed entrance. I come to a halt in front of the opened door, hesitating for a moment about going inside. This place doesn't look like any library I've ever seen. [...] Just inside the entrance a young man is sitting behind a counter. (Murakami 2005: 36)

The man behind the counter, the gatekeeper if you like, is Oshima, a hermaphrodite whose presence there seems to suggest that this library will be a place where definitions, like male and female, have been dissolved, a place where it might be possible to bridge or bypass certain distinction. Compared to George Kien, who at one point in *Auto-da-Fé* 'dreams of the end of woman kind' (Canetti 1984: 183), and Jorge, who rules an abbey into which women aren't even permitted, Oshima seems to be the gatekeeper of a library which at least accepts the notion of sexual difference.

When Kafka Tamura enters the library and walks among the stacks, his description of them is gardenesque.

> I go into the high-ceilinged stacks and wander among the shelves [...]. Magnificent thick beams run across the ceiling of the room and soft early-summer sunlight is shining through the open window, the chatter of birds in the garden filtering in. (Murakami 2005: 39)

The plum trees at the gate way, the well-manicured garden, the beams running overhead and the chatter of birds, all these details put me, at least, in mind of other libraries in literature: libraries which have been likened to forests or tended gardens. For instance, the library that appears in Jean Paul Sartre's *Nausea* is at one point described as follows:

> The reading room was almost empty. [...] It was as light as mist, almost unreal, all reddish; the setting sun was casting a reddish colour over the table reserved for women readers, the door, the spines of the books. For a second I had the delightful feeling that I was entering a thicket full of golden leaves. (Sartre 1965: 229)

And at the end of Borges's short story 'The Book of Sand', it is in the library that the protagonist hides the diabolical book, maintaining that 'the best place to hide a leaf is in the forest' (Borges 2001: 93).

This interplay between the natural world and the library is a theme that Murakami returns to regularly in *Kafka on the Shore*. It is interesting to note that before the protagonist is invited to live in the library he is taken to a place in the forest, as if time spent in the natural world will prepare him for his movement into the space of the library. A mood of initiation infuses Kafka Tamura's interface with the forest. Alternately, later in the novel, it seems that living in the library is an initiation for yet another excursion into the forest.

Related to this movement from the library to the forest and back again is the movement back and forth from the physical world to the world of the imagination. Indeed, not long after having entered the library Kafka Tamura states: 'As I relax on the sofa and gaze around the room a thought hits me: This is exactly the place I've been looking for all my life. A little hideaway in some sinkhole somewhere. I'd always thought of it as a secret, imaginary place, and can barely believe that it *actually exists*' (Murakami 2005: 39). A library space which was once an idea has grown in to a reality. Yet this movement of the library from being an imagined space to an actual space is somehow left unfinished, and the reader is left in some doubt as to the reality of the library's existence. What Murakami has done, in fact, is create a space in which, if I can quote Gaston Bachelard's *Poetics of Space*, 'the function of the real and the function of the unreal – are made to co-operate' (Bachelard 1969: xxxi). In this library, fact and fantasy, real and unreal, the actual and the imaginary coexist, and what's more, it is a place where, in Bachelard's words, 'the imagination is ceaselessly imagining and enriching itself with new images' (Bachelard 1969: xxxii). It is a place where the doors to daydreaming are unlocked (Bachelard 1969: 14).

When Kafka Tamura takes up his abode in the library, the library becomes his house and in becoming his house we can say of it what Gaston Bachelard says of houses. He writes:

> If I were asked to name the chief benefit of the house, I should say: the house shelters daydreaming, the house protects the dreamer, the house allows one to dream in peace. (Bachelard 1969: 6)

Murakami's library is a space in which dreams are closest to the surface. The library space in which Kafka Tamura resides becomes somewhere from which he is able to access remote regions where, as Bachelard writes 'memory and imagination remain associated, each one working for their mutual deepening' (Bachelard 1969: 5).

And it is a deepening without an emphasis on verticality. What is important is what we might liken to a signifying chain: from the station to the gateway, then down a path which winds through a garden, from the garden to a doorway, and then past a counter and into the library. There is no fortification around it, no complicated locks on the door. Porous, perforated and personal are adjectives that come to mind. After navigating through the dark rooms of Eco's labyrinthine abbey library and suffering with a monomaniac in Canetti's claustrophobic one, it is with some relief that the reader finds and explores Murakami's. We sense that we have found an invented library space where it is possible to think and dream freely.

### *Cartographers of Consciousness: the Shape and Composition of the Mind*

So having elaborated on these three invented libraries, I feel compelled to compare the narratives in which they appear to a story Baudelaire translated into French, known to English speakers as 'The Purloined Letter', in which, according to Lacan's reading of it, 'it is the letter and its detour which governs [...] [the character's] entrances and roles' (Lacan 2006: 21). In Eco's narrative, it is the forbidden book which governs the character's movements. Certain signifiers in the library space seem to have summoned the cast of characters, who, finding themselves in each other's company, manoeuvre around each other in a race to wrestle the said signifiers from their secret space: a hidden chamber in the centre of the library.

In *Auto-da-Fé* it is the library itself which dictates the movement of the characters, for it is the library to which the character's attach their dreams and desires, however shabby, base and extravagant their dreams and desires may be. And indeed, communication in and around Canetti's library space is constantly impeded by the degree to which all of Canetti's characters are so engrossed in their own ambitions that communication with anyone else, in any real sense, is impossible. Nor are Canetti's characters able to engage with the library space. Only Peter Kien is capable of seeing beyond the dollar value of his library; however, his vision is a narrow and fragile one, which is completely detached from reality. In this sense, by the end of the narrative, the library is entirely debased.

In *Kafka on the Shore* the library space also has a central role; however, Murakami's library is less centred than Eco's or Canetti's. Ownership of the library is not over emphasised, because the library is seen to belong to those who are in need of it. Murakami's library is also a transitory space; the characters of Murakami's fiction have a nomadic quality about them. Not for a moment are we given the impression that any of them are there to stay. They enter the library to be transformed and to play a part in the transformation of others. And such transformation is possible in Murakami's invented library, because in being a place where the protagonist feels safe, a place in which he can dream freely, it becomes possible to conduct soundings into the deepest recesses of the human organism where the symbolic has taken hold (Lacan 2006: 11). Lacan maintains that 'nothing, however deep into the bowels of the world a hand may shove it, will ever be hidden there, since another hand can retrieve it, and that what is hidden is never but what is *not in its place* [manque à sa place], as a call slip says of a volume mislaid in a library'(Lacan 2006: 17).

Umberto Eco writes that

> We have always been amazed by those Humanists of the fifteenth century who rediscovered lost manuscripts. Where did they find them? They found

> them in libraries: in libraries that served in part to conceal books, but also served to enable them to be found again. (Eco 2005: 8)

In this sense, libraries can be seen as not only mirrors of the world, as Eco's narrative suggests, but also as spaces that reflect the multiple regions of the psychic apparatus (Derrida 1996: 19). 'The whole structure of language that psychoanalytic experience discovers in the unconscious' (Lacan 2006: 413) can, I imagine, have a no more fitting model than the space of the library. Although I would be reluctant to limit any such model to just one library. The impenetrable library of the despot, the Apollonian library begging to be debased, and the inviting and tranquil library in which the mind is free to dream, all these libraries have corresponding structures in the language rich layers and regions of the conscious and unconscious mind.

And of what are these inner libraries composed? In answering this question I draw heavily on Gaston Bachelard's *Poetics of Space*. In these inner libraries 'creation takes place on the tenuous thread of the sentence'. These inner libraries are fibred spaces 'traversed by the simple impetus of words that have been experienced' (Bachelard 1969: xxiv). In these libraries of the interior images flow into lines of words carrying the imagination along with them, as though the imagination were creating nerve fibres (Bachelard 1969: xxiv). There are no doubt dead and redundant regions, but there are also regions alive with discovery and the true joy of creation. And consciousness, as well as being something we experience, is something we create.

## *References*

Bachelard, Gaston. *The Poetics of Space*. Boston: Beacon Press, 1969.

Batchelor, David. *Chromophobia*. London: Reaktion Books, 2000.

Borges, Jorge Luis. *The Book of Sand and Shakespeare's Memory*. London: Penguin Books, 2001.

Canetti, Elias. *Auto-da-Fé*. New York: Farrar, Straus and Giroux, 1984.

—. *Crowds and Power*. New York: Farrar, Straus and Giroux, 1984.

Derrida, Jacques. *Archive Fever: A Freudian Impression*. Chicago: University of Chicago Press, 1996.

Eco, Umberto. 'De Bibliotheca.' *Libraries*. Ed. Candida Höfer. London: Thames & Hudson, 2005.

—. *The Name of the Rose*. London: Secker & Warburg, 1983.

Lacan, Jacques. *Écrit*. New York: W. W. Norton & Company, 2006.

Murakami, Haruki. *Kafka on the Shore*. London: The Harvill Press, 2005.

Sartre, Jean-Paul. *Nausea*. London: Penguin Books, 1965.

Scianna, Ferdinando. 'Italy, 1984. Jorge Luis Borges.' *Writers: Photographs of Magnum Photos*. Ed. Jean-Claude Dubost. Paris: Éditions Pierre Terrail, 1998.

# Experiencing Kiefer's 'Scorched Earth' Landscapes: Acts of Re-enactment, Acquaintance, or Empathy?

R.A. Goodrich

In the following I set out to explore the experiential gap when we, as viewers, encounter a visual artwork. In particular, I shall focus upon Anselm Kiefer's large 'scorched earth' landscapes of the seventies onwards which critics, when recognizing his allusions to war, have had little compunction claiming represent the unrepresentable. Set against this apparently paradoxical state of affairs, I shall specifically analyse the extent to which our efforts to connect to the experience purportedly framed by the work or explored by the artist can be explained in cognate acts of consciousness – re-enactment, acquaintance, or empathy – on the part of an engaged viewer. Each act implies various 'embodied concepts' concepts, as George Lakoff claims, that do not 'exist independent of the bodily nature of any thinking beings and independent of their experience' (Lakoff 1987: 12; cf. Lakoff and Johnson 1999: 16–22). Each of these three acts of consciousness, he subsequently contends, extends 'beyond mere awareness of something' and is manifested by particular spatial metaphors (Lakoff and Johnson 1999: 11). In effect, re-enactment is manifested by the spatial metaphor of movement or motion, acquaintance by location, and empathy by projection. In terms of the spatial metaphor permeating this chapter, each section of it asks, how does experience bridge the distance between a viewer and an artifact?

Along with Joseph Beuys, Anselm Kiefer ranks amongst the best known and most provocative of German artists to have emerged in the generation following the collapse of the Nazi tyranny. Notwithstanding the intense political and cultural reactions to Kiefer's work (e.g. Huyssen 1989), it is the experiential assumptions of early critical response to his 'scorched earth' landscapes that shall be pursued here. The American reception to exhibitions of his work has included such damning responses as James Gardner's depiction of his 'truculent mediocrities' which only indicate that he 'has nothing to say' and 'says it badly' (Gardner 1989: 52). Or again, Arthur Danto finds his 'sludged and operatic fabrications' to be 'willfully obscure' and so 'spontaneously equates obscurity with profundity' in 'a farce of heavy symbolism' (Danto 1989: 26–27). Beneath such rejection is an appeal to a kind of experiential authenticity on the part of artist and spectator alike. This is made manifest when Danto accuses Kiefer of employing 'purely external' symbols, the kind we consult in library references rather than construct from lived experience, because 'his experience is merely that of an art student who has no life to speak of or draw upon' (Danto 1989: 27). As a result, Kiefer, unlike Beuys, ultimately fails in Danto's eyes to make works of art where it can be truly said 'one encountered one's deepest self in encountering them' (Danto 1989: 27).

Nonetheless his work is seen by many commentators on both sides of the Atlantic as one which mediates between a 'deeply traumatic' historical past and current viewers who (not unlike the artist himself) 'cannot fully know' it (Saltzman 1999: 2). Yet somehow, Lisa Saltzman assures us, the work 'bears witness and provides access' to that very history. For Saltzman, Kiefer's art 'probes, thematizes, and instantiates' an aesthetic if not an ethical 'dilemma' of 'how to represent the unrepresentable'. It is a dilemma or 'a confrontation made particularly acute by the knowledge that history, and particularly history after Auschwitz, can be encountered, grasped, and understood only through the acknowledgement of the very inaccessibility of its original occurrence and experience' (Saltzman 1999: 16).

Whatever else Saltzman might mean by unrepresentable experience in terms of her art-historical concerns, conceptions of unrepresentability can range widely. They may, for example, be exemplified by an inability or incapacity to exhibit, show, or depict an image of something or to make it appear; to use or serve it as a symbol of or a sample for something; to make it correspond or to be in part equivalent or analogous to something; or, finally, to bring it to mind or to impute something. The fact is that Kiefer is capable of representing facets of the Nazi past by way of his series of 'scorched earth' landscapes – beginning with such 1974 large-scaled, high-horizon, predominantly brown and black, mixed-media canvases as *Maikäfer flieg* (*Cockchafer, fly*), *Malerei der verbrannten Erde* (*Painting of*

*Scorched Earth*), *Märkische Heide* (*Brandenburg Heath*), and *Nero malt* (*Nero Paints*). Scrawled across the horizon of the first named, as Mark Rosenthal notes (Rosenthal 1987: 32), is a German children's adaptation at the war's end of an old nursery rhyme:

> Cockchafer, fly
> Father's at war,
> Mother's in Pomerania,
> Pomerania's burnt down.

On reflection, however, it seems Saltzman is not analyzing unrepresentability as such. Rather, the conception is tied to an inability by either artist or viewer to re-capture the original experience to be represented (though whether artistically, morally, phenomenologically, or all three tends to be left open in her monograph). Even amongst less vitriolic critics, the elusiveness of re-capturing experience is explicitly acknowledged. Mark Stevens, for instance, also asserts: 'No artist who is not himself a survivor can pretend to look the crime full in the face' (Stevens 1988: 29). Yet the question remains: why should the inability to re-capture an original experience on the part of artist and viewer alike be regarded as problematic? That artworks are or should only be rendered aesthetically acceptable if they can represent the content of an original non-artistic experience retrievable by both artist and viewer would seem to be an unsustainable stipulation. It threatens to eliminate fictional or allegorical subject-matter altogether.

Perhaps this kind of appeal to original experience can be evidenced elsewhere. By analogy, it is not uncommon for many of us to hear old soldiers recounting tales of their survival during battle or incarceration. Suddenly, in the midst of these recollections, we find them asserting to those who have never experienced the events in question, 'if you have not supped from the same cup, you cannot know the taste'. Listeners might well agree with this sentiment in acknowledgement that, no matter how many times we have been in pain or have felt fear, we cannot literally possess or experience the pain or fear of others. Our empathy with the old soldiers, it might be contended, therefore does not lie in assuming their pain or fear. In other words, perhaps it lies less in any duplication of their past experiences and more in our capacity to re-enact them.

So, before pursuing the question of empathy further, what does it mean to say that we can re-enact the past experiences of others? What does Kiefer himself mean by his often-cited comment, 'I do not identify with Nero or Hitler, but I have to re-enact what they did just a little bit in order to understand the madness' (cited in Stevens 1988: 28 and Danto 1989: 26)? Drawing upon R.G. Collingwood's controversial account of the process of re-enactment when one has 'no direct or empirical knowledge of [...] facts [...] no transmitted or testimonial knowledge of them' (Collingwood 1993: 282), we need to envisage the situation with which our old soldiers were dealing and then see for ourselves as if we were the soldiers in question what possible alternatives those circumstances raised and what reasons there would be for choosing one course of action over another. By so apprehending their pain or fear, from Collingwood's perspective, we are not referring to the 'immediate experience' or 'immediate feeling' of 'having or undergoing the pain' or the fear (Collingwood 1993: 291). Rather, we are referring to a self-conscious awareness of our experience as 'an experience of a determinate kind: an act, and an act of thought which has arisen in a certain way, and has a certain cognitive character' Collingwood 1993: 291). In such a case of re-thinking, or re-constructing the history of, an experience, we are not performing an act of recollection in the way our informants, the old soldiers, are 'in memory the past is a mere spectacle, in history it is re-enacted in present thought' (Collingwood 1993: 293). Moreover, their memory may bring with it recollections of the immediacy, the context, of the experience on that first occasion – emotions, feelings, moods, sensations, other thoughts – that cannot form part of our re-enactment at another time in another context.

Returning to the dark, massive landscapes of Anselm Kiefer, to talk of re-enacting the experience of the artist (rather than that which he intends to represent) is to talk of how we, as more or less artistically or critically informed viewers, might re-construct the painterly choices he made in the pursuit of an artistic idea. Yet Kiefer, who remains 'belligerently monosyllabic' as Gardner expresses it (Gardner 1989: 52), is an artist who pointedly avoids discussing his work in detail. As he remarked in a 1987 interview:

> I can only make my feelings, thoughts, and will in the paintings. I make them as precise as I can and then after that [...] you decide what the pictures are and what I am. (cited in Biro 1998: 267, n. 51, and Arasse 2001: 83)

Hence, one viewer, Simon Schama, arguably the most influential English-speaking intellectual initially to contextualize the earlier work of Kiefer, succinctly contends: 'As a work from 1974 made explicit (*Malen = Verbrennen* [*Painting = Burning*]), Kiefer came to think of his painting as an aggressive re-enactment of historical destruction, literally as a "burning". So where Romantic art reiterated the sentimental celebration of native landscapes, his art did what history did: it burned them' (Schama 1995: 83).

Yet reconstructing Kiefer's painterly choices may not enable us to revive the immediate experience of undertaking such work, reputedly 'despair' (Mooney 1998: 101). It therefore does not function as the process that underpins the frequently heard appeal of how we cannot appreciate an artefact in the final analysis without directly or immediately experiencing the work for ourselves. Re-enactment, as depicted above, does not so much result in 'an attempt to investigate and expound eternal verities concerning the nature of [...] Art' as in the effort 'to reach, by thinking, the solution of certain problems arising out of the situation in which artists find themselves' (Collingwood 1938: vi).

Let us return to the analogy of listening to old soldiers recalling their pain or fear. Whether this analogy collapses because it appears to have centred upon apprehending another's experience rather than upon evaluating it remains a persistent problem in accounting for our consciousness of art. As Richard Shusterman recently expresses it: 'If defining art as immediate aesthetic experience directs us profitably toward the core of art's value, such experience itself cannot take us [...] far [...] in justifying our evaluative verdicts. [...] the immediacy of aesthetic experience is in itself mute [...] evaluations of art require that [...] experience be filled out or anchored' (Shusterman 2003: 406).

Generations of philosophers have upheld Lisa Saltzman's appeal to first-person, first-hand experience, but usually as a precondition of its assessment. For example, we find David Hume reminding his readers that, without practice, we simply cannot discern the 'distinguishing species of each quality', let alone assign it 'suitable praise or blame', since we must allow people 'to acquire experience in those objects' so judged (Hume 1965 [1757]: 13). Skilled judgement, he continues, is to be accorded to those 'accustomed to see, and examine, and weigh the several performances, admired in different ages and nations' (Hume

1965 [1757]: 14). A not dissimilar appeal to experience can be found in Immanuel Kant's third *Kritik* when he comments that '[w]e cannot press upon others by the aid of any reasons or fundamental propositions our judgment that a coat, a house, or a flower is beautiful. People wish to submit the object to their own eyes, as if satisfaction in it depended on sensation' (Kant 1951 [1790]: 50, §8).

Why this should be so, it seems, may be found in the supposed fact that mere persuasion of others without recourse to their experience more often than not only ends with inferences about how we, not they, personally regard the object of our judgement. For both Hume and Kant, experiential corroboration is a minimum if not necessary requirement for securing aesthetic evaluation.

Leaving aside these seminal statements of European Enlightenment thinking and turning to aesthetic debate within the past generation or so, we need only consider some of the ways in which the appeal to experience has retained its centrality. In an extended notebook entry of February 1948, Ludwig Wittgenstein ponders how we might best describe the understanding of music. He concludes that it can neither be 'a process that accompanies hearing' nor a set of sensations (Wittgenstein 1967b: sec. 162 & 165). Yet understanding a musical phrase or a musical theme is tantamount to 'experiencing something whilst we hear it', sometimes revealed by expressive gestures, sometimes by how one 'plays, or hums, the piece', sometimes by the comparisons one draws (Wittgenstein 1980: 69 & 70). Wittgenstein has no hesitation here acknowledging the embodied nature of our understanding of the arts. Ultimately, for him, understanding this particular phrase or that particular theme is anchored to one's general understanding of music. It is because '*this* concept of understanding has some kinship with other concepts of experience' and all the more so because one is 'at home with the special conceptual world that belongs to these situations' (Wittgenstein 1967b: sec. 165). To take one example of what a theme might mean, writes Wittgenstein, '[h]ere it's as if a conclusion were being drawn, here as if something were being confirmed, *this* is like an answer to what was said before' (Wittgenstein 1967b: sec. 175), so that understanding the theme 'presupposes a familiarity with inferences, with confirmation, with answers' (Wittgenstein 1967b: sec. 175). It is just as if one were, in this case, explaining or teaching the meaning of an argument or demonstrating how Kiefer deploys the outline of a transparent palette in his 1974 'scorched earth' landscapes.

The generation since Wittgenstein has largely adopted the older, evaluative trend noted above. Mary Mothersill, for instance, explicitly asserts that any verbalized act of judgement of taste 'presupposes, through the avowal that it implicates, first-personal knowledge of the object judged' (Mothersill 1984: 54; cf. Scruton 1974: 54).[1] Indeed, she believes this intuition is revealed with anyone who praises or condemns a canvas – say, Kiefer's massive 1982 acrylic, emulsion and straw *Nürnberg* [*Nuremberg*] – but who, it later emerges, has not seen it. Such a situation leaves us feeling as if we have been significantly deceived. Like the American critics surveyed at the beginning of this chapter, it is personal experience which acts as a precondition of subsequent appraisal or as if recognition of something as an artwork is at one and the same time the recognition of its artistic value. The latter interpretation makes any appeal to experience one which cannot be a neutral affair. So, compared with Wittgenstein, Mothersill has shifted our attention from the kinds of experience promoting an understanding of artifacts to an evaluation of them. Richard Wollheim summarises this kind of appeal to experience under that 'well-entrenched principle' which he dubs the 'acquaintance principle', one 'which insists that judgments of aesthetic value, unlike judgments of moral knowledge, must be based on first-hand experience of their objects and are not, except within very narrow limits, transmissible from one person to another' (Wollheim 1980: 233).

Although Wollheim does not elaborate upon these limits, he does assert that the notion of experience correlated with evaluation needs to be construed in 'an unnaturally broad way' (Wollheim 1980: 232). Its breadth apparently ranges across 'mental' responses to artifacts that

---

[1] Scruton 1974, when pursuing the contention that certain aesthetic descriptions are non-descriptive in the epistemological sense that they do not express beliefs but experiences, claims that its initiating intuition should be understood as follows: 'In matters of aesthetic judgement, you have to see for yourself [...] we use a perceptual verb ("to see") in expressing this opinion.' Why this should be so, he continues, is because 'what you have to "see" is not a property: your knowledge that an aesthetic feature is "in" the object is given by the *same* criteria that show that you see it. To see sadness in the music and to know that the music is sad are one and the same thing'.

may be 'immediate or protracted, spontaneous or cultivated' (Wollheim 1980: 233).

Before questioning Wollheim's statement of the 'acquaintance principle' further, the question still remains whether acquaintance must always be tied to evaluation. Let us now re-contextualize the pre-occupation with the appeal to first-person, first-hand experience by briefly turning to the epistemological debate surrounding acts of acquaintance. Paul Hirst, for example, though focused upon the nature of education in the arts, concedes in accordance with the philosophical tradition outlined above that coming to know an artifact 'necessitates [...] direct experience of the work' and that this 'existential element' or 'such personal experiences' play a central role in cases of acquaintance (Hirst 1974: 154). Whilst acknowledging Bertrand Russell as his source for the notions of knowledge by acquaintance and by description, Hirst never clarifies which of Russell's several attempts to elucidate their technical distinction informs his own argument.

When initially defining the term 'acquaintance', Russell directs our attention to the subject's relationship to certain objects or things: 'I am *acquainted* with an object when I have a direct cognitive relation to that object, i.e., when I am directly aware of the object itself. When I speak of a cognitive relation here, I do not mean the sort of relation which constitutes judgement, but the sort which constitutes presentation' (Russell 1953 [1911]: 197).

However, how are we to understand a 'presentation' in this passage? Is Russell referring to the presentation of some thing, namely, an object, such as Kiefer's 1982 *Die Meistersinger* canvas? Or to something presented, namely, some aspect of an object, such as the pieces of straw or cardboard applied to the same canvas? Or is Russell referring to both? Furthermore, how is awareness more cognitive than, say, affective? Perhaps greater clarity can be found in Russell's *The Problems of Philosophy*, with its more popularized version of the distinction. Here, acquaintance is explicitly formulated in terms of complete, direct awareness. It is an awareness that is both immediate and incorrigible, yet not necessarily asserted verbally or mentally, because it is 'without the intermediary of any process of inference or any knowledge of truths' (Russell 1912: 73). In other words, Russell separates our experiential encounter from any evaluation arising from it.

Clearly, we have departed from the kind of acquaintance discussed in the previous section. So, from Russell's perspective, with what may we be acquainted? He stipulates four kinds or modes of acquaintance in explicitly embodied terms and initially by way of spatial metaphors. The four modes involve the sensations of our 'outer' five senses; the introspection of our 'inner' sense, namely, desires, feelings, and thoughts; memories of the former two; and, possibly, the self 'as that which is aware of things or has desires towards things' (Russell 1912: 80–81). In addition to these four existential particularities, Russell asserts that we also have acquaintance with what he calls '*universals*, that is to say, general ideas' (destruction and fragility to take two instances typically attributed to the work of Kiefer) (Russell 1912: 81). However, for Russell here, all knowledge, be it of physical objects and other minds or persons, is said to be knowledge by description. The importance of the latter for Russell lies, firstly, in being 'ultimately reducible to [...] what is known by acquaintance', and, secondly, in allowing us 'to pass beyond the limits of our private experience' (Russell 1912: 92).

Common usage upheld by Hirst unhesitatingly departs from Russell's definitions. In effect, Hirst maintains without argument that knowing a person or an artifact, for example, is classifiable as knowledge by acquaintance. However, Russell generally assigns knowing persons to knowledge by description, although such knowledge in turn supposedly bases itself upon '*constituents with which we are acquainted*' (Russell 1912: 91). What might justify such a reversal of Russell? First of all, we might conclude that acquaintance, being a dyadic relationship between a subject and some object, or a subject and his or her consciousness of some object, involves a certain degree of discrimination as exemplified by Russell's case of 'learning to be acquainted with whiteness' by 'a process of abstraction' Russell 1912: 159). But such a process of discrimination involving analysis and comparison is tantamount to an act of assessment, evaluation, or judgement. To adopt this line of argument (culled from Hicks 1919: 162 & 169) implies, by Russell's own prescription, that an act of acquaintance presupposes knowledge by description. In the second place, we might well be influenced by the fact that the term 'acquaintance' is indiscriminately applied to contexts comprising both persons and things. At times, we talk of having knowledge of some person or thing – Anselm Kiefer himself or his ‘scorched earth’ landscapes – which is more than mere recognition but less than intimacy. At times, too, we talk of making ourselves familiar or becoming familiar with Kiefer or his ‘scorched earth’ landscapes. At other times, we even talk of being accustomed to, informed about, or cognizant of Kiefer or his landscapes or conscious or informed *that* Kiefer or his ‘scorched earth’ landscapes are, say, pre-occupied with the Nazi era. Hence, it could be maintained that we would scarcely be straining common usage to speak of persons or artifacts as an instance of knowledge by acquaintance.

However, both lines of attack in support of our common use of ‘acquaintance’ are open to objection. The first part of our argument makes the illicit assumption that the means thought necessary to convey an instance of acquaintance, namely, by a process of abstraction, logically implies something about the procedure necessary to have or experience acquaintance of that instance (as first noted by Broad 1919: 209). Consider the situation where someone reports his or her visual sensation of devastation when first seeing Kiefer’s *Nürnberg* by way of describing comparable colours, vistas, states of mind, and the like. Yet giving such a report need not imply anything about the means necessary for him or her to have acquaintance with devastation. Furthermore, even if our acquaintance with some object were always preceded by various acts of discrimination – comparison, analysis, and so forth – this need not imply that we are ultimately unable to be acquainted with that object. Someone may judge that the lowering of the gallery lights intensifies his or her perception of the darkness of Kiefer’s other ‘scorched earth’ landscapes. Again, this perceptual discrimination does not automatically prevent his or her capacity to be acquainted with their darkness. The second part of our counter-argument on behalf of common usage confuses the concept of acquaintance with that of knowledge by acquaintance (cf. Broad 1919: 206). In so far as knowledge involves true justified belief or judgement, acquaintance may well be a constituent of knowledge without itself being knowledge, just as oxygen may well be a constituent of life without itself being life. In other words, acquaintance, as direct awareness, is not a judgement and therefore lies outside the realm of truth and falsity. Even if we are correct in wanting to dismiss knowledge by acquaintance from a consideration of the consciousness peculiar to the arts, our grounds for doing so should not rest upon a misconception of what Russell was attempting to argue.

Let us now return to Wollheim's articulation of the acquaintance principle introduced at the end of the third section of this chapter. Of late, the acquaintance principle, at least in its bald form, has been attacked on several fronts by Paisley Livingston and by Malcolm Budd (see Livingston 2003 and Budd 2003). In summary, they question its plausibility in a number of ways. Three sets of questions posed should suffice for our purposes. First of all, is first-hand experience to be denied in cases where what is directly experienced is a reproduction – be it electronic, or holographic, or photographic – of that object? To pursue the example of Kiefer's *Nürnberg*, what is it that makes monochromatic reproductions barely the size of a postage stamp appropriate on some occasions in ways that, say, forgeries which more precisely reproduce texture, materials, dimensions, colour, and the like are inappropriate? Indeed, in a case where a work no longer exists such as the 1981 version of *Dein goldenes Haar, Margarethe* (*Your Golden Hair, Margarete*), we would have little choice but to refer to its reproduction.[2] The same holds if a work were damaged beyond repair – a not inconceivable fate awaiting many of Kiefer's works given the decomposable nature of the materials he employs. Needless to add, there may be no uniform means of determining what kinds of reproduction satisfy the features of the original given the sheer variability of purposes served by them.

Secondly, is it invariably true, as we found Wollheim claiming, that 'first-hand experience' is basically not 'transmissible from one person to another' (Wollheim 1980: 233)? Imagine the case of a viewer of the 1982 work *Nürnberg* whose initial evaluation of it based itself upon a perception of that work. But nowadays the viewer is no longer able to recall its appearance over time, although he or she still retains a memory of the original judgement. Does such a case now demonstrate an instance of non-transmissibility of experience within the one person? In what way is such a person where acquaintance is no longer operative different from another who maintains the same judgement, but one overtly based upon the testimony of others? Alternatively expressed, is it in the nature of an aesthetic judgement

[2] See, e.g., the small black-and-white reproduction in Arasse 2001: 108 (with palette outline in straw).

that it cannot allow for its formation through, say, the testimony of others? Indeed, without the combination of reproductions and testimonies, students of art the world over would be excluded from learning to understand, let alone to compare and appraise, works of art.

Thirdly, does it obviously hold that our judgements or evaluations are entirely grounded upon the perceptual or directly experienced features, character, or properties of an artifact? Do they not equally derive from referential, semantic, or contextual aspects such as the very title of the work or a knowledge of the work's allusions, genre, influences, provenance, and style? If so, then does what constitutes a direct perceptual experience of the *Nürnberg*, for all the variations accompanying an actual sighting of it, also lie within a complex background of knowledge often drawn from the testimony of others? We need only contrast the differences in background reverberations for viewers with and without a knowledge that, for example, the city of Nuremberg is the setting of Hitler's favourite opera, Wagner's *Die Meistersinger*, as well as the site of Nazi rallies of the 'thirties and of war trials of the late 'forties calling Nazi leaders to account.

Does the appeal to acquaintance as a means of bridging the gap between viewer and experience represented, let alone between viewer and artist – with all its attendant epistemological complexities – ignore a seeming conundrum that might be more explicitly revealed by turning to the cognate activity of empathy? Whereas the notion of sympathy – feeling *with* or *Mitfühlung* – has of course had a long history of discussion since classical antiquity, empathy – feeling *into* or *Einfühlung* – emerged as a distinctive if related term of analysis in aesthetic debate in the late nineteenth century. Typically, to preserve the difference between the two, critics tend to resort to etymological stipulation. As an example of someone who attempts to counter this tendency, Lauren Wispé postulates the distinction between both kinds of feeling as follows:

> In empathy one substitutes oneself for the other person; in sympathy one substitutes others for oneself. To know what something would be like for the other person is empathy. To know what it would be like to be that person is sympathy. In empathy one acts 'as if' one were the other person. [...] The object of empathy is understanding. The object of sympathy is the other

person's well-being. In sum, empathy is a way of knowing; sympathy is a way of relating. (Wispé 1991: 80)[3]

Returning to our earlier anecdote, were we to respond to those old soldiers recalling their pain or fear in battle by indicating that we know just how they felt, it is clear that their adage, 'If you have not supped from the same cup, you cannot know the taste', rejects our claim to empathy. In other words, if they are asserting that we cannot know how they felt, they are simultaneously making a claim about how we feel. Yet, if it is true that someone, say, Jack, cannot know how another, say, Jill, feels (or thinks or perceives), then does it not also follow that Jack cannot know that Jill does not know how Jack feels in the first place? The conundrum of rejecting empathy, as Ramsey McNabb observes (McNabb 2005), turns upon one of two claims: either it is impossible for one person to know the exact feelings of another, or it is impossible for one person to understand with any adequacy the feelings of another.

If Jack adheres to the first of the claims, it fast becomes obvious that he has no justification for knowing the exact feelings of others short of pretending to possess telepathy. Having no access to another's consciousness, another's mental state, cuts both ways. If Jack insists upon the second of the claims, then the question is not so much whether the sum of each unique person's feelings could ever be identical with the sum of another's. Rather, the possibility that some feelings can be had by both parties – be they our old soldiers and their listeners or a viewer and an artist – remains open. Jack has no way of proving that Jill cannot have the same feeling about a specific matter since that again would require direct access to the mind of Jill, let alone to a judgement of Jill's understanding of or even imaginings about particular feelings had by Jack. Ludwig Wittgenstein expresses the difficulty in these terms: 'Could someone understand the word "pain", who has never felt pain? – Is experience to teach me whether this is so or not? – And if we say "A man could not imagine pain without having sometime felt it" – how do we know? How can it be decided whether it is true?' (Wittgenstein 1967a: part 1, sec. 315)

3 Cf. Nussbaum 2001: 327–335, on empathy as the imaginative reconstruction of another's experience.

Yet, at the beginning of Wittgenstein's enquiry which ends with the questions just cited, he recognizes that '[i]f language is to be a means of communication there must be agreement not only in definitions but also [...] in judgments [...] [which] is partly determined by a certain constancy [...]' (Wittgenstein 1967a: part 1, sec. 242). In other words, our ability to share concepts, our capacity to respond in similar ways, both making language and communication possible, suggests that we can empathize or imagine how another feels, providing, as Susan Feagin reminds us (Feagin 1996: 95–96), the other, or the mode of representing the other, is the object of our empathy. Doubt only emerges if Jack is aware of certain facts or occurrences not known to Jill, facts or occurrences influencing what Jack feels in a particular case. As McNabb concludes, 'disproving the [very] possibility of empathy requires empathy' (McNabb 2005). Hence, it would seem that empathy can bridge the space between viewer and artist, between a viewer and the apparent experience represented or expressed by a work.

In conclusion, this chapter has critically examined the extent to which our efforts to connect to the original experience purportedly represented or expressed by an artwork or attributed to an artist can be explained in terms of cognate acts of re-enactment, acquaintance, or empathy on the part of an engaged spectator. Influential appeals to re-enactment tend to see such acts by viewers in largely cognitive or intellectual terms removed from the immediacy of experience. Similarly, seminal discussions of acquaintance quickly become embroiled with contestable issues over the experiential and epistemological prerequisites for formulating aesthetic judgements or evaluations of artworks. It therefore appears that we, as spectators, are left with empathy for all its seemingly paradoxical impossibility as our fundamental means of bridging the experiential gap with an artist and his work.[4]

*References*

Arasse, Daniel. *Anselm Kiefer*. Transl. Mary Whittall. New York: Harry N. Abrams, 2001.

[4] Particular thanks are owed to Peter Leech (Otago) and Elissa Goodrich (Melbourne) for their incisive comments about earlier drafts of this chapter.

Biro, Matthew. *Anselm Kiefer and the Philosophy of Martin Heidegger.* Cambridge: Cambridge University Press, 1998.

Broad, C.D. 'Is There "Knowledge by Acquaintance"?' *The Aristotelian Society* suppl. Vol. 2 (1919): 206–220.

Budd, M.J. 'The Acquaintance Principle.' *The British Journal of Aesthetics* 43/4 (October 2003): 386–392.

Collingwood, R.G. *The Principles of Art.* Oxford: Clarendon Press, 1938.

—. 'History as Re-enactment of Past Experience.' [1936] *The Idea of History with Lectures 1926–1928.* Ed. W.J. van der Dussen, rev. ed. Oxford: Clarendon Press, 1993. 282–302.

Danto, A.C. 'Anselm Kiefer (Museum of Modern Art, New York).' *The Nation* 248/1 (2nd January 1989): 26–28.

Feagin, S.L. *Reading with Feeling: The Aesthetics of Appreciation.* Ithaca: Cornell University Press, 1996.

Gardner, James. 'Anselm Kiefer at MOMA.' *National Review* 41/4 (10th March 1989): 52–54.

Hicks, G.D. 'Is There "Knowledge by Acquaintance"?' *The Aristotelian Society* suppl. vol. 2 (1919): 159–178.

Hirst, P.H. 'Literature and the fine arts as a unique form of knowledge.' *Knowledge and the Curriculum: A Collection of Philosophical Papers.* London: Routledge & Kegan Paul, 1974: 152–164.

Hume, David. 'Of the Standard of Taste.' [1757] *The Standard of Taste and Other Essays.* Ed. J.W. Lenz. Indianapolis: The Bobbs-Merrill Co., 1965. 3–24.

Huyssen, Andreas. 'Anselm Kiefer: The Terror of History, the Temptation of Myth.' *October* vol. 48 (Spring 1989): 25–45.

Kant, Immanuel. *Critique of Judgment.* [1790] Transl. J.H. Bernard. New York: Hafner Press, 1951.

Lakoff, George. *Women, Fire, and Dangerous Things: What Categories Reveal about the Mind.* Chicago: The University of Chicago Press, 1987.

— and Mark Johnson. *Philosophy in the Flesh: The Embodied Mind and Its Challenge to Western Philosophy.* New York: Basic Books, 1999.

Livingston, Paisley. 'On an Apparent Truism in Aesthetics.' *The British Journal of Aesthetics* 43/3 (July 2003): 260–278.

McNabb, Ramsey. 'The Paradox of Empathy.' *Philosophy Now* 52 (August/September 2005) at: http://www.philosophynow.org/issue52/52mcnabb.htm.

Mooney, Ted. 'Anselm Kiefer ('Nürnberg' Painting).' *Artforum International* 36/8 (April 1998): 100–101.

Mothersill, Mary. *Beauty Restored.* Oxford: Oxford University Press, 1984.

Nussbaum, M.C. *Upheavals of Thought: The Intelligence of Emotions.* New York: Cambridge University Press, 2001.

Rosenthal, Mark. *Anselm Kiefer: Itinerary of the Exhibition: The Art Institute of Chicago and Others.* Chicago: Prestel Verlag, 1987.

Russell, B.A.W. 'Knowledge by Acquaintance and Knowledge by Description.' [1911] *Mysticism and Logic and other essays.* Harmondsworth: Penguin Books, 1953. 197–218.

—. *The Problems of Philosophy.* London: Williams & Norgate, 1912.

Saltzman, Lisa. *Anselm Kiefer and Art after Auschwitz.* New York: Cambridge University Press, 1999.

Schama, Simon. *Landscape and Memory.* New York: Alfred A. Knopf, 1995.

Scruton, R.V. *Art and Imagination: A Study in the Philosophy of Mind.* London: Methuen & Co., 1974.

Shusterman, Richard. 'Pragmatism Between Aesthetic Experience and Aesthetic Education.' *Studies in Philosophy and Education* 22/5 (September 2003): 403–412.

Stevens, Mark. 'A Poet of the Apocalypse.' *The New Republic* 198/6 (8th February 1988): 27–30.

Wispé, Lauren. *The Psychology of Sympathy.* New York: Plenum Press, 1991.

Wittgenstein, L.J.J. *Culture and Value.* Ed. G.H. von Wright and Heikki Nyman, transl. Peter Winch, 2nd ed. Oxford: Basil Blackwell, 1980.

—. *Philosophical Investigations.* Ed. G.E.M. Anscombe and Rush Rhees, transl. G.E.M. Anscombe, 3rd ed. Oxford: Basil Blackwell, 1967a.

—. *Zettel.* Ed. G.E.M. Anscombe and G.H. von Wright, transl. G.E.M. Anscombe. Oxford: Basil Blackwell, 1967b.

Wollheim, R.A. 'Art and evaluation.' *Art and Its Objects: With Six Supplementary Essays.* 2nd ed. Cambridge: Cambridge University Press, 1980: 227–240.

# CONSTRUCTIONS OF SELF IN SPACE

# Fractured Urban Memories

Dirk de Bruyn

> There is a pursuit of knowledge foreign to language and founded upon visual communication, demanding a development of the optical mind, and depending on perception in the original and deeper sense of the word.
>
> Stan Brakhage

*Fractured Urban Memories* consisted of the screening of the 55 minute 2 Screen 16 mm film: 'Experiments'[1] with *live* interventions/manipulations of the image with shadow-play, masking, coloured and polarising filters and the movement of the projectors themselves so that the images at times overlapped and at times sat side by side. Each film had a separate soundtrack with often competing optical sound, whose volume was also manipulated during the performance. As the interacting artist I was constantly visible to the audience during these operations as the audience members were seated to the side of the projectors in front of a white wall on which the two moving image streams interacted with each other. (Please note that henceforth, for the sake of brevity and simplicity, within the body of this paper this event will be referred to as 'The Performance'.)

To profile a trace of the original work the following extract from a short review is offered, written in 1982 for the *Melbourne Times* newspaper by Rod Bishop: 'Projected on two screens, with two separate soundtracks, the always exceptional, and occasionally brilliant, photographic images are enhanced by de Bruyn's rigorous control over a wide variety of experimental techniques. Without overindulging in any of them, de Bruyn uses animation, optical illusions, time lapse, solarization, hand tinting, flash frames, refilming and flicker effects, accompanied by a dense atmosphere of word puns,

---

[1] Experiments (1981, 2 screen, 55 mins, 16mm, Optical Sound).

dialogue, primal screams, music and even recycled and letraseted soundtracks. By setting experiments entirely within his Moonee Ponds house, de Bruyn creates such a complex sense of claustrophobia, the spectator, while recognizing the staid, conservative trappings of urban Melbourne, is presented with the sort of art neurosis more commonly found in mega-cities like New York.'

Trying critically to frame your own art can be an uncomfortable, some may say, a narcissistic, undertaking. It seems reminiscent of Merleau-Ponty's 'double sensation' (Merleau-Ponty 1962: 106) of both touching and being touched, of touching one hand with the other hand. It is a situation that itself echoes or speaks to this ambiguous status of our bodies as both object and subject suggesting a phenomenological reflexivity worthy of further scrutiny inside this text.

The restaging of this performance in 2006 for the conference *On Space* at the University of Otago afforded an opportunity to discuss a number of issues, including the nature of an expanded cinema space and how the experimental is marked as source for contemporary ideas about interactivity, database and modularity in digital media and VJ culture. It also enables a case to be made for experimental film as a unique self-reflexive genre/category of film and video production. I will also take the opportunity to discuss how, through its sampling, looping and repetition, through the physicality of optical effects like flash, flicker and afterimage the experimental can explore the visual representation of dissociation and neurosis within this space.

And there is also the issue of memory. Do such structures as are explored in this performance have any relationship to how memory works? Replaying excerpts 25 years on from their first presentation and through a live re-manipulation of the image another layer of memory and loss is added to the mix of this precariously recorded and re-presented neurotic moving visual field.

### *The Direct*

The performance was tendered in lieu of a spoken presentation, in part to stress Edward S. Small's contention in *Direct Theory* (Small 1994) that experimental cinema is a stand-alone genre of filmmaking that bypasses the written and spoken word to state its theory directly, through the construction of the film itself. Small's direct theory contends that experimental film as a genre does its theorizing intrinsically and directly through its methods of assembly, through the making of the work itself: 'A type of film and video production that I contend does not function mainly as (fictive) popular entertainment nor as (documentary) information. Its major function is rather to theorize upon its own substance by reflecting back on its own intrinsic semiotic system(s)' (Small 1994: 5). Small also argues that experimental films illustrate Rudolph Arnheim's concept of visual thinking. Arnheim's critical work examined visual perception through a lens of Gestalt psychology. His approach insists that the unique perceptual processes set up in watching Small's 'direct' films should be considered as thinking in themselves. 'My contention is that cognitive operations called thinking are not the privilege of mental processes above and beyond perception but the essential ingredient of perception itself' (Arnheim 1969: 13). This approaches a phenomenological view that the story that unfolds in such experimental work is a narrative about perception itself.

Within experimental film's theorizing, its visual thinking, Small argues, the written becomes redundant, unnecessary. At the same time, it needs to be said, neither does the irony escape me, nor should it the reader, that this argument about optical thinking is now communicated and formulated in textual form. Is this the ambivalent space of Merleau-Ponty's janus-like 'double sensation' that the performance itself opens up? The dynamic of this ambiguity, this jockeying for primacy between the spoken with the written and the seen, is of itself signifying something of interest, a phenomenological tension that shifts between the visible and invisible, the seer and the seen which Sobchack extends to a 'doubled' cinematic space of the viewer and the viewed: 'Cinematic vision, then, is never monocular, is always doubled, is always the vision of two viewing subjects materially and consciously inhabiting, signifying, and sharing a world in a manner at once universal and particular, a world that is mutually visible but hermeneutically negotiable' (Sobchack 1992: 24).

The self-reflexive mode of direct theory that Small articulates emphasizes the structural, architectural emphasis of a non-narrative cinema. 'Instead of foregrounding recognizable characters and narratives, the (experimental) theoretical films foreground the mechanical, chemical, perceptual, and conceptual structures that underlie the theatrical film experience in general' (MacDonald 1988: 2–3). Small lists eight characteristics that help define the genre:

reflexivity, mental imagery, innovative use of technology, brevity, avoidance of verbal language and narrative structure, economic and artistic autonomy.

Small links the concept of self-reflexivity to Russian formalist Viktor Shklovsky's concept of 'baring the device' (Small 1994: 55–6). Shklovsky also theorizes about defamiliarization or ostranenie that aims to disorientate the viewer so as he/she can experience an object anew. This is evident in The Performance where the repetitions and flicker assault are designed to disorientate and reproduce direct aspects of the experience of a migrant's estrangement, the negative impact of his/her surroundings, which are experienced as a direct, unprocessed, assault on the senses. 'The purpose of art is to impart the sensation of things as they are perceived and not as they are known. The technique of art is to make objects "unfamiliar", to make forms difficult, to increase the difficulty and length of perception because the process of perception is an aesthetic end in itself and must be prolonged' (Shklovsky 1965: 12).

While reflexivity is but one of our eight generic characteristics, it can effectively bond the others to reveal experimental film/video's unique function as a kind of extant, manifest, immediate theory: direct theory (Small 1994: 22). In its most effective form *reflexivity* has to do with the notion that the work transmits an awareness of itself. The work exhibits self-awareness. It talks about itself within itself. It is not a question of suspending disbelief while you are immersed in a subjective spectacle. There is an 'open-ness', a direct-ness, a clear awareness that you are watching a film or video, and that this awareness is creatively enmeshed, intrinsic to the experience of the work. Structure and content speak to each other.

*Fractured Urban Memories* operates through its seriality, its repetitions, contemplations and manipulations. Such works are composed of non-narrative structures where the story, at the very least, is not fore-grounded. Non-linear ways of presenting information are 'played' with, elements can be repeated, layered, sampled, disrupted, erased, fragmented and mixed up in unusual ways, often inspired by the strategies of collage. 'When experimental film/video does deal with narrative [...]. It typically presents fragmented narratives that tend to confound the conventions of classic continuity; thus these classic conventions stand either parodied or de-constructed' (Small 1994: 21). Though 'more than a film', the transitory performance meets all but one of direct theory's eight criteria (brevity) marking it for this dubious honour of inclusion in this marginalized non-narrative/non-linear tradition. Numerous repetitive texts inserted and didactically spoken in the performance assert, remind the audience that 'this is not a narrative film'.

Such structures as those embraced in The Performance's image stream are also intrinsic to many new or digital media works. Malcolm Le Grice, leading Structuralist film-maker and foundation member of the London Film Maker's Co-op, argues for the continued relevance for experimental film's reflexive theorizing in new media. In *Experimental Film in the Digital Age* he notes the digital enabling of a non-linear cinema that finds its beginnings in experimental film: 'The tradition of experimental film and video has already provided the basis for exploring these concerns (of non-linearity) not as a response to the new technology but as a consequence of artistic reflection on the human condition' (Le Grice 2001: 296).

The observation of the non-narrative tradition passing through the 'experimental' to new media art, offers a further rationale for re-presenting this performance 25 years after its initial publication. This re-insertion also occurs at a time of the saturation of new media or digital art within contemporary arts practice matched by a parallel erasure and forgetting of experimental film art within Australian art exhibition.

Lev Manovich's championing of Dziga Vertov's *Man With a Movie Camera* (16mm 1929 Russia), widely considered as an originating work for experimental filmmakers, is also significant in confirming this experimental/digital media convergence. Manovich creatively appropriates frames from this film as foundation topography for the prologue to his seminal 2001 text *The Language of New Media* and further recycles these images as interface, to signpost different sections of the text in that book. Manovich's branding of *Man With a Movie Camera* as meta-film fortifies this experimental/new media connection: 'Vertov is able to achieve something that new media artists still have to learn – how to merge database and narrative into a new form' (Manovich 2001: 243).

The tradition of experimental cinema, also evident in digital media, is part of an ongoing project that uses innovative audiovisual techniques in the construction of a personal space. Such a reflexive use of media is primed to directly articulate ideas concerning the construction of the self, the body and memory. According to Small, Eisenstein's montage, which can be regarded as the collage of the moving image, has been a dominant strategy within this project: 'Now why should the cinema follow forms of theatre and painting rather than the methodology of language, which allow wholly new concepts of ideas to arise from the combination of two concrete denotations of two concrete objects?' (Eisenstein 1949: 60).

Just as the re-combination of images one after the other, a la Eisenstein, is so essential to a language of the moving image; within the two screen film the appearance of images side by side, or one on top of the other opens up this theatrical/audience space even more along a room's x and y co-ordinates. There is also the body of the performer inserted in this space on which the image registers (along an z axis, if you like), leaving a masked trace on the wall.

The three dimensional space that is opened up in The Performance with the insertion of this z axis, stresses the centrality of the body. The artist's body is continually present and the film itself, the visibility of the projectors, in its reflexive baring the device of the viewing experience interrogates the 'existential and embodied act of viewing. We must reflexively and reflectively take possession of our vision and make it visible' (Sobchack 1992: 54).

### *Denial and Loss*

Within such a reflective and reflexive space what terminated, denied or erased past emotional state can be communicated? This is a question that the performance tries to deal with. I have come to realise that the task of how you represent a loss, or a denied trauma has been part of an ongoing project in my film and video work. It is an issue I was taken back to during the performance and in the discussion that followed. Part of the original aim of *Experiments* was to document a response to the emptied space of suburbia, the suppressed anger at residing within its barracks. It was to act as a directed resistance to being in this space.

Melbourne suburbia, the site and subject of the first performance of *Fractured Urban Memories*, can be identified as a terminal for European trauma and loss. It is a layered collage of migration, a field of identities torn from somewhere else trying to forget, to begin again. For the migrant there is also a resistance to belonging that needs to be negotiated. To be identified as a 'new Australian' had a pejorative edge as an empty officially sanctioned term or identity that placed those so named on an outside masquerading as an inside.

Aspects of experimental film reflect features of this migrant experience as a marginalized 'do it yourself' practice, in its resistance to storytelling and its embrace of new media. As such it appeals as a practice through which to articulate the nature of a marooned migrant experience, a neutered identity and the impact of being on the outside.

There is also a political appeal to the experimental that lies in an attempt to distil ideas of *resistance* and *direct action* so much part of the student anti-war movement of the sixties and to the terrorism of Baader-Meinhof, the Weathermen, the Black Panthers or the Red Brigade. Unfortunately, through its aesthetic and theoretical transformation of these concepts experimental film has, like punk, marked a diminished political impact for these ideas into a less deadly, more manageable marginalized social practice. In its quest to develop a new and innovative direct language of independence and opposition the focus of direct action has been transferred inwards onto the materiality and structure of the work rather than directly at the political power structures operating in society. There remains, however, an inscribed trace on the body of film of a politics of dissent. This can be understood as a process of withdrawal from the political arena, a containment of the originating impulse, as a kind of repression or forgetting.

This dynamic of denial, of the forgotten or unspoken can also be found in the relationship between new media and experimental film in Australia. The new comes in and erases the old and the old becomes invisible. It is not seen or recognized for its foundation work. Often the history or story of art that new media art in Australia does build on and relate to is one that comes from somewhere else, from outside, from overseas. This is a persisting colonial mind-set. This attitude is a progeny of that dominant 'alpha' Australian tradition of erasure, whose founding 'terra nullius' gesture was the raising of the British flag by the first fleet on 'virgin' soil, as if there was nothing here. And so, there is this recurring process of denial that can mark the migrant's assimilation into their new world. A kind of tension of expression, a dis-association, where the emotional residue, an internal architecture, a body trace, of past experiences remains, but the content and history of those stories are rendered irrelevant. I have also suggested how some of these features are evident in aspects of how experimental film is received and placed in Australian conditions.

Such a focus on structure rather than content in areas of experimental film can be categorized as a formalist or modernist approach to the use of moving images. As its extreme manifestation (and this can also hold for new media work) this formalism maintains that art has no content and that, in fact, the structure is the content. I would contend that such an approach, such a dynamic could also productively be read as content *denied.* With such an interpretation we bring the hidden, the underground, the unspeakable, the invisible into play and we bring into focus those structures, those architectures, those games that hold those denied secrets in our culture, including how we keep in our bodies the traumas of the past. Such a highly modernist formalist approach to moving image making can be read as 'baring the device' of denial.

Adorno and Ramadonovic discern a similar trace of suppressed emotion in their historical analysis of modernist painting and the holocaust. Within such an interpretation Greenberg's naming of modernism's project as the stripping back to an absolute of appearance becomes an empty vision constructed from the mechanisms of erasure rather than of revelation. This idea of *content denied* manifests itself in the perceived relationship between the holocaust and modernism. Ramadanovic (2002) shows that within the immediate period after the Second World War, the holocaust impacted a triple convergent break with war-trauma itself, with history and with representation and resulted in modernist art, in response, being emptied of content.

Adorno has pointed out that before Auschwitz death was conceived as an individual experience: 'Auschwitz confirmed the philosopheme of pure identity as death' (Adorno 1973: 363). His concept of 'total indifference' depicted 'what it meant to survive'. In the Holocaust '[t]he killing itself was abstracted in such a way that it would not make sense to represent victims anymore as individual subjects, since this would miss the point of the horror as "administrative measure"' (Ramadanovic 2002: 238). In the period after the the Second World War 'art was encouraged to show a representational break with the past that indicated the presence of a historical caesura marking the so-called post-war period' (Ramadanovic 2002: 235). This was a forgetting designed to ensure that history not re-write itself, that the politics of the past would not be important in constructing the future. Perhaps, remembering Aragon's comments about the appropriation of surrealism by advertising after the First World War: 'Dwight McDonald said it well: to describe was to accept the unacceptable' (Guilbaut 1983: 197). The concern was that the representation of the horrors of war would be received and reprocessed into kitsch. In respect for the holocaust's far-reaching impact such an outcome needed to be avoided.

Adorno's *Negative Dialectics* in which he argued that historical trauma could no longer be responsibly depicted, reinforced such concerns. His 'absolute negativity' 'shatters the basis on which speculative metaphysical thought would be reconciled with experience' (Adorno 1973: 362). As a result thought becomes dissociated from experience.

This climate heralded a move into abstraction 'avant-garde art became an art of obliteration, an act of erasure' (Guilbaut 1983: 181). 'At the very least, one could infer that we had now entered an age when the unspeakable of history could only be addressed by the unspeakability of art' (Ramadanovic 2002: 238). Looking to move beyond abstract expressionism, this project of erasure is encapsulated by the anecdote of Robert Rauschenberg's performative erasure of a Willem de Kooning drawing in his *Erased de Kooning Drawing* (1953), 'the product of one month and forty erasers spent rubbing out the crayon, grease pencil, and ink markings of a drawing that de Kooning had good-naturedly given him as a gift' (Solomon 1997: 236). Here was a momentum of technique that pushed painting so far that 'by the late 1950's Yves Klein and others would start thinking in terms of the death of painting' (Ramadanovic 2002: 236). For Greenberg, this was seen more as a stripping back to an 'absolute of appearance' in art.

The political imperative that 'the past was not to play an important role in the new', seems strangely reminiscent in the idea of the new Australian, new media in Australia as well as the historical sublimation of experimental film. This imperative pushed art into the abstract and neatly circumvented any need to deal with the holocaust in the same way that surrealism had critically focused and tried to process the mutilation residue of the First World War. 'Helpful in the fabrication of this historical illusion was the flight of art into radical abstraction, since this stylistic move was itself a representational breach with not only fascism but with many of the avant-garde movements that had been in effect during the pre-war period, among them expressionism, fauvism, surrealism and cubism' (Ramadanovic 2002: 235).

How can an art practice, so apparently 'directly' empathetic to the impact of the Holocaust, become so unhinged as to also, in its forgetting impart the opposite impact, create a reactionary total indifference? Is not such a moving from any representation of the monstrous, ironically, the mechanism and impetus of denial?

How can this happen? Kroker (2003: 204) points out that with all technologies, the bias in communication flips its figure/ground once that technological realm is played out, exhausted. Perhaps the bias of an empathic response to the holocaust flipping into denial is the natural result of a progression, exhaustion through technique. Such a reversal is also articulated in McLuhan's (1988) concept of the tetrad, his four laws of media, the fourth of which suggests that every innovation when pushed to its accepted limit converts to its opposite form.

I will conclude this discussion with an anecdote designed to keep this ambivalence, this janus-faced dilemma, hanging in mid air. I finish with a parable from Witold Gombrowicz's *Cosmos* (1994) in which the protagonist while out for a walk comes across a large rock that must be circumvented. Having done this he is so incensed that he has let this monstrosity affect his actions so that he decides to return and walk right over the top of it instead. Having completed this erasure of his original denial he realizes that he has let this boulder affect him doubly.

So which is it? Is it about facing the trauma, the monstrous, and the unpalatable directly or moving on? It seems, inevitably, that the trace remains, no matter what action you take. Film that talks back directly and materially about itself, exhibits a self awareness, seems like an important tool in communicating such inconsistencies. Twenty-five years tells me that history is important and that it still resides in the space, the architecture of my body, its cluster of senses – a rich honour.

*References*

Adorno, T. *Negative Dialectics*. New York: Continuum, 1973.

Eisenstein, Sergei. *Film Form.* Transl. Jay Leyda. New York: Harvest, 1949.

Gombrowicz, W. *Cosmos and Pornografia: Two Novels.* London: Grove Press, 1994 (Re-Issue).

Le Grice, Malcolm. *Experimental Film in the Digital Age.* London: BFI Publishing, 2001.

MacDonald, Scott. *A Critical Cinema.* Berkeley: University of California, 1988.

Manovich, Lev. *The Language of New Media.* Cambridge, Mass; London: MIT Press, 2001.

Merleau-Ponty, Maurice. *Phenomenology of Perception.* Athens: Ohio University Press, 1962.

Shklovsky, Victor. 'Art as Technique.' *Russian Formalist Criticism: Four Essays.* Edited by Lee T. Lemon and Marion J. Reis. Lincoln: University of Nebraska Press, 1965. 3–24.

Sobchack, Vivian. *The Address of the Eye: A Phenomenology of Film Experience.* New Jersey: Princeton University Press, 1992.

# Curating Curiosity: Wonder's Colonial Phenomenology

Khadija Z Carroll

> In places like universities, where everyone talks too rationally, it is necessary for a kind of enchanter to appear.
>
> Joseph Beuys

> We may be this — but to what end?
>
> Julie Gough

> No words can adequately capture the meaning of these objects. Unspoken in their purpose, to seek to comprehend them is to lend an ear to other voices.
>
> Phillippe Peltier

In the Pacific islands on one of his voyages of discovery Captain Cook looked at the extraordinary Tongan *Tapa* (coconut-fibre cloth) and remarked: 'curious'. New Zealand paddles he noted were 'curiously stained'. Parkinson, a member of his crew, said of fish hooks that they were 'curiously carved', and the body of the Maori 'curiously tattooed'. Marquesian head ornaments he called 'a curious fillet of shell work decorated with feathers'; Marquesian diadems were similarly 'curious' (cited in Thomas 1991: 130). A sailor on board the *Endeavour* wrote: '[S]everal of [the Marquesans] had caps very curiously wrought in shapes not inelegant, and composed of feathers interspersed with spangles of mother of pearl, that looked very gay and were very becoming' (1991: 130).

Traders in curiosity, the Cook party expected at least a return gaze, or evidence of a common desire on the part of indigenous peoples to help them make sense of and integrate the new. But as they sailed down the Australian east coast in 1770 for the first time the Aborigines they encountered looked straight through them – as if they did not exist, as if refusing to accept that an object such as a ship must appear 'so remarkable [...] to people who ha[d] never seen one' (Thomas 2003: 111). To the Australian Aborigines, the *Endeavour* was part of an incommensurable new cosmology.

Cook's party soon found to their disappointment that the Aborigines had no 'abundance of superfluities' to exchange.[1] Nor did the Aborigines invest any of their rich imagination in absorbing the meaning of the gifts they had been offered. They abandoned the fine neck-cloths and hair-ribbons they had been given and just walked away. What few items they possessed they had made themselves, and, like all refined design, these were objects of simple utility. Take, for example, a spearthrower (*woomera*) collected in 1923 from the Kimberley region in northwest Australia by E. Clement and later sold to the Peabody Museum.[2] The archive of drawings Clement made show that this spearthrower has two sides (Figure 1). Clement's drawings indicate a second dimension to the flattened one-dimensional object shown in the museum display case. The neat double of front and back in his drawings correlates with the received notion of a 'hybrid' as a mixture of two different things. Yet this object has at least four operative surfaces and might therefore be called a 'recombinant hybrid' – a hybrid that resists even this two-fold classification. The spearthrower allowed a hunter to throw his spear three times further (up to an extraordinary 180m). The *cadjie-cadjie* incisions on the *woomera*'s reverse side also indicate the object's involvement in an entirely different space – the *woomera* becomes the source of that most immaterial of arts in the sacred dances of a Corroboree when a boomerang is rubbed along these incisions to make music. What Clement doesn't mention in his notes is that, as well as percussion instrument and lever in spear hunting, a spearthrower may also be used as a mixing tray for pigments or tobacco, a utensil to make fire by friction and a wood-working tool.

---

[1] An observation from the archive of Captain Arthur Phillip's first settlement in Sydney (cited in Clendinnen 2003: 32). Clendinnen brilliantly reconstructs the Aboriginal Australians' strategic diplomacy and negotiation with the first settlers. In my own reading of the incommensurable nature of Australian space to the Enlightenment explorers I in no sense wish to perpetuate the view that there was no thoughtful agency in the negotiations that Clendinnen masterfully reconstructs from the settlers' statements.

[2] Archive of the Peabody Museum of Archeology and Ethnology at Harvard University.

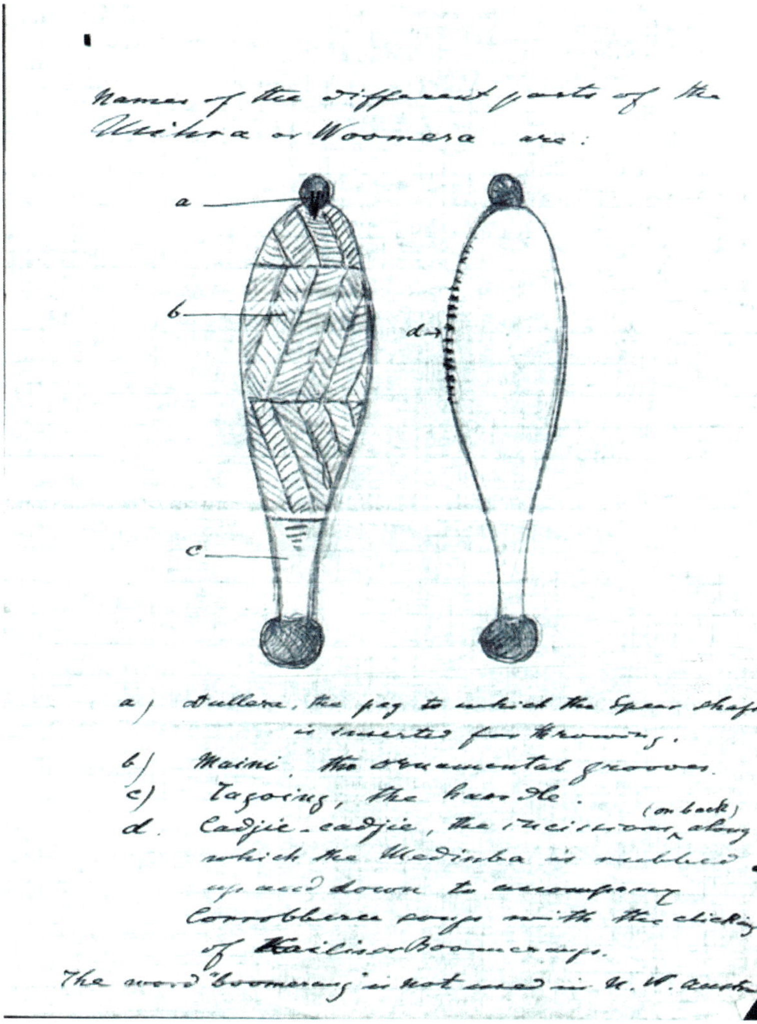


Figure 1: Archive drawing by E. Clement
Archive of the Peabody Museum of Archeology and Ethnology at Harvard University

What should we call an object whose 'thingness' appears so removed from our own understanding of a commodity?[3] As Clendinnen observes, the issue of naming arises in the case of the spearthrower and many other objects in ways '[o]ften symptomatic of a wider incomprehension' (Clendinnen 2003: 103). Just as artifacts become artworks when their maker is known, so objects become subjects for the museum when they can be viewed as animate. Colonial acts of (re-)naming subsumed the singular within a schema or preconceived whole on the basis of the operative category of curiosity. The category of curiosity is to be seen as distinct from wonder. For, rather than experiencing the sublime wonder that a new system presents, the colonizers brought the new they encountered on their voyages of discovery within parameters of understanding for which 'curiosity' served as a key term.

In analyzing the methods of display of the museum it becomes clear that taxonomy is by no means neutral, but is rather indebted to prevailing ideology. The system of taxonomy that the academy and the museum still follow today is descendant from the universal taxonomy issuing from the work of the Enlightenment botanist, physician and zoologist Carl von Linné. The burgeoning interest in science and geographical expansion in Linné's day meant that the older aristocratic museums with their small rooms gave way to the 'cabinets of curiosities' in public display in the museums of the eighteenth century, thereby also making a new system of classification necessary.[4] With the discovery of the New World and new trade

---

[3] The Australian response to commodities contrasts with Marcel Mauss' theory of gift exchange and extra-domestic trade. Yet Mauss also argues that objects in the Melanesian *kula* trade are not inanimate, indifferent things, but objects that are coveted – named, attributed personality, a history and even mythological status to the extent that people may be named after them. It is perhaps the particular material culture of Australia that leads to the conflation of the perceiver and the perceived in encounters where a ritualistic or religious dimension is present. Cf. Mauss 1967.

[4] The English word 'museum' derives from the Greek *mouseion*, meaning 'seat of the Muses'. In Boston this high ideal was used as a front for the scandalous theatre. The Boston Museum offered its visitors the educative entertainment of viewing curiosities before going to see the theatre discreetly hidden in the same building. The Boston Museum collection was replete with the crowd pleasing faux mermaid skeletons were then donated to Harvard in 1866 to begin the Peabody Museum. In a sense the Boston and Peabody Museums' foundation in spectacle and visual deception – in favour of engaging people's sense of curiosity – throws another light on current Marxist critiques of the 'Disneyfication' of museums.

routes, artifacts rapidly found their way into these eclectic and encyclopedic eighteenth century cabinets. 'Cook's journals had set the style and established the taste for dramatic doings in exotic places which could be elevated to science', Inga Clendinnen writes, 'by the inclusion of observations of curiosities encountered along the way: of birds, plants, animals and savages, usually in that order' (2003: 103).

Bronwen Douglas has argued that the scientific information that flowed back into Europe as a result of Cook's voyages contributed to a decline of neoclassical idealism in art and science and so helped bring about an important shift in taxonomy. By the end of the eighteenth century a new romantic sensibility in art and literature had triumphed, and a 'biologization' of the human sciences was promoting a new 'evolutionist' cosmology. While the earlier humanists thought of 'race' as a venerable variety within a divine whole, the scientific Enlightenment reconceived 'race' as a set of permanent physical differences between human groups that were passed on by hereditary to later generations (cf. Douglas 1999). Faced, therefore, with the unknown, wonder and curiosity came to mark quite different responses to the new. Curiosity, unlike wonder, does not illicit helplessness. There is no remainder once you have accounted for curiosity, while wonder entails an unaccountable remainder. The term I use for this unaccountability, this inability to measure, compare or even comprehend, is incommensurability. Like antinomy, incommensurability signals a contradiction or incompatibility in thought arising from the attempt to apply to the ideas of the reason relations which are appropriate only to the concepts of experience.[5] In the colonial discourse from Captain Cook onward, curiosity and wonder shape the terms of response to incommensurability.

While curiosity and wonder often overlap in the same discourse, in the colonial context curiosity views novelty as something commensurable when wonder is lacking. I would like to propose that 'curiosity', which has been markedly omitted from inquiries in aesthetic philosophy, makes an interesting contribution to our understanding of the colonial encounter. In considering what wonder

[5] A history of curiosity is yet to be written. Hepburn 1984 offers an insightful account of wonder from Plato via Aquinas to Leopardi by way of Francis Bacon, Adam Smith, Shakespeare, Kant, Heidegger, Levi-Strauss, Heinrich Heine and Wittgenstein. Antinomy is a term in the Kantian philosophy, while 'incommensurability' is treated as a term in Deleuze 1968.

might contribute to a colonial phenomenology I also treat hybridity as a possible strategy of adaptation to the new (and thus also survival) for indigenous culture. This is a story I tell in interwoven narratives: for more than two hundred years the *woomera* speaks of encounters with Aboriginal art in different keys. The different encounters work in combination and recombination. They do not constitute a linear narrative leading directly to the responses of artists today. Rather, the historiography of naming and classifying operates in at least three temporal registers: the indigenous taxonomy of objects; the collection enterprise of the explorers, and the contemporary status of 'Australian art'. I understand hybridity as the equivalent in art to the notion of recombination in physics and genetics where characteristics are combined differently from the way they functioned in a previous entity or self. 'Recombinant hybridity', I argue, offers a way to reconcile oneself to a lack of consistency. I chart this hybridity as it shifts from a purely conceptual category to one I find to be the governing practice of some Aboriginal artists today.[6] In the reconciliation of what is ancient to a new future, hybridity has an element of modernity not usually conceded to indigenous understanding.

The first contact with New Holland was contact with an incommensurable culture, and the incommensurability of two worlds was manifest at the moment of exchange. It is in these exchanges that vastly different temporalities collided and in a sense updated or 'recombinated' each other. The language that is used to talk about those objects seen in the first encounter sheds light on this cross-cultural incommensurability. Bernard Smith has suggested that 'to say that an object was "curious" was to express an interest in it without passing an aesthetic judgment' (cf. Thomas 2003: 130). In this view curiosity – the Cook party's obsessive incantation in responses to indigenous culture – stood for an absence of aesthetic judgment.

To specify further the relationship of curiosity to this absence of aesthetic judgment we may look at Wittgenstein's notebooks from 1916 where he writes about the aesthetic experience as wonder. The aesthetic moment, he observes, is one of wonder that the world exists,

[6] Brook Andrew, Julie Gough, Fiona Foley, and Christian Bumbarra Thompson to name just a few.

'that what exists, exists'.[7] He notes that we do not wonder at *how* the world is, only *that* it is. Wittgenstein refines this idea in the *Tractatus logico-philosophicus* where the term 'aesthetic wonder' shifts to become a central concern with 'the mystical'. Another way to translate these relational and spatial terms is that we wonder *at* the world but not *about* how it functions. Wittgenstein himself spatializes this claim in the next proposition in the *Tractatus*:

> 6.45 The contemplation of the world sub specie aeterni is its contemplation as a limited whole. The feeling that the world is a limited whole is the mystical feeling.
> *Die Anschauung der Welt sub specie aeterni ist ihre Anschauung als – begrenztes – Ganzes. Das Gefühl der Welt als ein begrenztes Ganzes ist das Mystische.*

A return to his *Notebooks* from 1916 shows that Wittgenstein explicates wonder *sub specie aeterni* in relation to the artwork. He lays out the claims in the following set of propositions:

> The usual way of looking at things sees objects as it were from the midst of them, the view sub specie aeternitatis, from outside.
> *Die gewöhnliche Betrachtungsweise sieht die Gegenstände gleichsam aus ihrer Mitte, die Betrachtungen sub specie aeternitatis von ausserhalb.*
>
> In such a way that they have the whole world as background.
> *So dass sie die ganze Welt als Hintergrund haben.*
>
> Is it perhaps in this view the object is seen together with space and time instead of in space and time?
> *Ist es etwa das, dass sie den Gegenstand mit Raum und Zeit sieht statt in Raum und Zeit?*
>
> Each thing modifies the whole logical world, the whole of logical space, so to speak.
> *Jedes Ding bedingt die ganze logische Welt, sozusagen den ganzen logischen Raum.*
>
> The thing seen sub specie aeternitatis is the thing seen together with the whole logical space.
> *Das Ding sub specie aeternitatis gesehen ist das Ding mit dem ganzen logischen Raum gesehen.* (Wittgenstein 1961: 85)

[7] Wittgenstein 1961: 86: *'Nicht wie die Welt ist, ist das Mystische, sondern dass sie ist. Das künstlerische Wunder ist, dass es die Welt gibt. Dass es das gibt, was es gibt.'*

Since it is a discipline's practice to legitimate a theoretical claim by providing an example, I will attempt to draw Wittgenstein back to the time of exploration, when the Europeans wondered at the world beyond their dominion. In Wittgenstein's terms, not being *in* this new world beyond Europe before the explorations, the Europeans perceived it as a wonder *sub specie aeternitatis* from outside. *That* a world existed outside the maps made by the ancient world was cause for a most terrible wonder.[8] It would only be later, *amongst* the contents of the New World, that the Europeans would become curious *about* the New World. That is why the first encounter is mysterious, because of a shift from a view *sub specie aeternitatis* to one *in* the time and space.

The literary genre of fantastic voyages and utopias attests to the European's wonder at the thought of the New World. In 1676 Gabriel de Foigny, for instance, in his novel about the as yet undiscovered southern land, cast Australians as blissfully self-sufficient hermaphrodites.[9] While the eroticization of the Aboriginals in his utopia had the Swiss priest defrocked, it could be interesting to ponder to what extent these fantasies of the New World determined the way material was later collected and constructed as evidence for the Europeans of their fantasies of the first encounters.

Just as the genres of utopia and fantastic voyages were in fact a thinly veiled critique of European society, Captain Cook also composed a Rousseauean panegyric about the Australian Aborigines:

> From what I have said of the Natives of New Holland they may appear to some to be the most wretched people upon the Earth, but in reality they are far more happy than we Europeans; being wholly unacquainted not only with the superfluous but the necessary Conveniencies so much sought after in Europe, they are happy in not knowing the use of them. They live in a Tranquility which is not disturb'd by the Inequality of Condition: The Earth and sea of their own accord furnishes them with all things necessary for life, they covet not Magnificent Houses, Household-stuff &c, they live in a warm and fine Climate and enjoy a very wholesome Air, so that they have very little need of

---

[8] How enormous the shock at sailing off the map, at going beyond the authorial dominion of the ancients, was only hinted at in Greenblatt 1991. Coleridge is cited dizzyingly: for 'outness' is but the feeling of otherness rendered intuitive, or alterity visually represented.

[9] Cf. de Foigny 1993. The first (1676) edition in French and the English translation published the following year are held in Houghton Library, Harvard University.

> Clothing and this they seem to be fully sensible of, for many to whome we gave Cloth & c. to, left it carelessly upon the Sea beach and in the woods as a thing they had no manor of use for. In short they seem'd to set no value on any thing we gave them, nor would they ever part with any thing of their own for nay one article we could offer them; this in my opinion argues that they think themselves provided with all the necessarys of Life and that they have no superfluities. (Cited in Thomas 2003: 128)

This reflection on the state of nature may be the only one in Cook's diaries, yet its causal explanation takes stock of their thwarted attempts at exchange with the Aborigines.[10] It renders evident the curious and rational nature that underlies Cook's firm grip on his purpose: to ascertain whether there was a southern continent, to map it, and to explore the possibility of trade. The experience of wonder, with all its attendant threats at disarming the individual of their expectations, seems at odds with this scientific purpose.

The relation of wonder to judgment can also be reconstructed on the basis of the diaries and drawings made on the Cook voyages. On his third voyage Cook stops at the southern most part of Australia, the island of Tasmania. He writes of the people there that were described to him as they appeared on shore after he had left:

> Many of the children had fine features, and were thought pretty; but of the persons of the women, especially those advanced in years, a less favorable report was made. However, some of the Gentlemen belonging to the *Discovery*, I was told, paid their addresses, and made liberal offers of presents, which were rejected with great distain; whether from a sense of virtue, or the fear of displeasing their men, I shall not pretend to determine. (Cf. Nordyke 1999: 9)

Cook's restraint in passing judgment on the women is heightened by his care to report what was told to him by Lieutenant King. Like Montaigne in his essay on cannibalism, the virtues of the other culture are reported with an awareness of an inability to pass judgment on cross-cultural matters.[11] In Montaigne's comparison of the indigenous people in Brazil to the ancient Greeks, commensurability is

---

[10] I think here of Karl Marx, 'The fetishism of commodities and the secret thereof', in *Capital*, Volume One, Section 4, where the example of Robinson Crusoe's island is used to read the mysterious nature of commodities and the social character of the labor that produces them. Cf. Marx 2003.

[11] It is not clear whether a lack of empathy, or an excess of incommensurability is the reason for this. Montaigne 1958: 151–159.

undermined by the act of cannibalism. With great curiosity Montaigne interviews the 'king' of the tribe. He is impressed by how he measures his army, showing Montaigne the size of the space it would take to hold approximately five thousand men. All this is 'discovered' through gestures to the ground, in a process of *showing* not *telling*, because, as Montaigne reports, his interpreter follows 'my meaning so badly, and was so hindered by his stupidity in taking in my ideas' (Montaigne 1958: 159). Montaigne finds the tribal leader's authority inscribed in the land when he asks whether his authority expires with war. He is told that as the cannibal moves between the villages that depend on him, his subjects make paths through the underbrush so he can travel comfortably through the jungle.[12] The tour de force self-reflexivity in Montaigne's cross-cultural encounter however comes in the final line of his essay: 'All this [means of spatial measurement etc.] is not too bad – but what's the use? They don't wear breeches' (Montaigne 1958: 159). Even the short, experimental form of Montaigne's essays reflects the limits he perceives in his own understanding. Judgment of another culture gives way to incomprehension and wonder. Though Montaigne may know something of their cultural practices, cannibalism remains incommensurable to the European's perspective.

The incommensurability of behavioural rules in inter-racial experiences of contact has a long history in Australia (cf. Mulvaney 1989). The attempt to assimilate the Aboriginals into an entirely British system of law and exchange is illustrated sixty years after Cook in the Proclamation Board, a schematic series of paintings made to be hung on trees (Figure 2). The British crown had proclaimed Australia unknown country, *terra incognita,* and thus free to be taken. The Proclamation Board illustrates the hypothetical assimilation of 'the savage' under British law. Within each scene the narrative painting feigns an equality through mirroring: when read from left to right the teleology runs from savage crime to civilized punishment. The inscription on the back of the board says it was made to advertise law *visually* what could not be communicated verbally.[13] Syntactically

[12] I would put 'the cannibal' and 'the jungle' in parenthesis to signpost my unease with the lack of a more specific name that could shift these entities from general exotic categories to a real example.

[13] In his proposal of this drawing to Governor Arthur, George Frankland writes: 'I have lately had an opportunity of ascertaining that the aboriginal Natives of Van

compact, such visual code was a dysfunctional strategy. A surveyor named George Frankland had seen Aboriginal paintings and thought they might therefore be educated by visual means. Yet soon after in 1830 the black war spelled the genocide of Tasmanian Aborigines. The two-dimensional representation of hanging literally hung on the premise of universal legibility and the rigidity of the colonial system gave way to an incommensurability of sign systems.

What is wonder in relation to curiosity, which sees itself as within a world it is able to know? The German Enlightenment philosopher Immanuel Kant distinguishes in his terminology between *Verwunderung* and *Bewunderung*. The difference in the prefixes of the root *Wunder* (wonder, literally) is significant. As in *Verstand* (understanding) the *Ver-* prefix in German indicates a displacement. *Ver-wunderung* is then a displacement of *Wunder*, and in place of wonder colonial phenomenology produced curiosity. An object is sublime, Kant writes, provided it does not *verwundern* as a 'novelty

---

Diemen's Land are in the habit of representing events by drawings on the bark of trees, and that the march of a certain party of Europeans, over a country before unfrequented by us was found a short time afterwards drawn with charcoal on a piece of bark, by a tribe of natives who been observed attentively watching their movements – the carts, the Bullocks, the men were distinctly represented, according to the exact number that really existed. In the absence of all successful communication with these unfortunate people, with whose language we are totally unacquainted, it has occurred to me that it might be possible through the Medium of this newly discovered faculty, to impart to them to a certain extent, the real wishes of the Government towards them, and I have accordingly sketched a series of groups of figures, in which I have endeavoured to represent in a manner as simple and as well adapted to their supposed ideas, as possible, the actual state of things/or rather the origin of the present state/and the desired termination of Hostilities. The proposal which I venture to make is that if your Excellency approves of the drawings, they should be multiplied, and being made on more durable materials, should be fastened to trees in those remote Situations where the Natives are most likely to see them. It is, at best but an experiment, but as it will be attested by neither expense, nor inconvenience, your Excellency may perhaps consider it worth trying.' *Archives of Tasmania* LSD 17/1: 23. Surveyor George Frankland to Governor Arthur , 4 February 1829. I thank Julie Gough for bringing this letter to my attention.

exceeding expectation' can.[14] Instead, *Bewunderung* is wonder or admiration that 'does not cease when the novelty wears off'.[15]

In light of Kant's telling distinction the word 'curious' is not neutral the way 'interesting' and 'peculiar' might be similarly common in the description of customs, occurrences and artifacts. These words certainly stand for the incommensurable in the first contact with the other in the colonies. However, the supplement to description in this supposedly 'neutral' term may betray some intent in the curious gaze. The political implications over time, furthermore, should lead to a rereading of colonial curiosity.[16]

Note how 'curiosity' both denotes the paddle, whistle, body, and describes the state of mind when looking at these. This confusion in terminology is possibly a strategy of the curious to deflect attention from their *Verwunderung*, the inability to grasp matched with a desire to control. If there was no aesthetic judgment in the curious apprehension of an artifact then there was certainly interest of a particular sort in the curiosity.[17] It is useful to reiterate that curiosity is distinct in this sense from wonder, which certainly has, though not exclusively, an aesthetic disinterestedness.

What kind of observation is curiosity then, in these curious accounts, these accounts of curiosities? These accounts were instrumentalized, the curiosity was satisfied and the curios were taken back to England, and some were bought by the Peabody Museum. A curious order of events in which the cabinet of curiosities in Europe became institutionalized by the modern state and the museum made

---

[14] As Thomas writes (2003: 124): 'New Holland was not the southern continent that had been sought, but it was a land replete with entirely new things. It appeared arid but was, for an emerging discipline obsessed with novelty, nevertheless a paradise.'

[15] Cf. Kant 1957: *Critique of Aesthetic Judgment* §29, General Remark; also §58. See also Critique of Teleological Judgment §1; and *Critique of Practical Reason*, Conclusion.

[16] Notably Kant is writing about nature when he explicates sublime and the disinterested view. Nature does not have interests; therefore it is well suited to the kind of viewing that Kant takes pleasure in. What is at stake for anthropologists such as Fred Myers is to deconstruct the art-culture system of the 'free and creative' fine arts that assume to transcend and critique 'use value'. What underlies this seemingly irreconcilable antagonism between the utilitarian and the aesthetic object is in part at least the basis of aesthetic philosophy. Cf. Myers 2002.

[17] Nicholas Thomas has taken on Bernard Smith's claim that to say an object was 'curious' was to express an interest in it without passing an aesthetic judgment (Thomas, 1991: 130).

the fetish for the curiosity into the science of taxonomy. Considering the colonial apparatus at work, can we still speak of curiosity as if it is a quality in the object that makes the viewer want to collect it?

The conflation of subject and object in the terms 'wonder', 'curiosity' and also 'marvel' is difficult to avoid. The experience of visual pleasure becoming transferred onto the object that gives pleasure is inscribed in these words. For instance 'I wonder' at 'the curiosity' – these words are used to describe both the visible object and the receptive experience internal to the subject. Stephen Greenblatt commits this conflation throughout when he describes wonder as the power 'to convey an arresting sense of uniqueness' (Greenblatt 1991: 1).[18] 'The power to convey' has the puzzling implication of both the subject's readiness to wonder and to project that power of wonder onto the object *and* the object or wonder's active agency in this process.[19]

The inability to account for novelty also underlies the urge to order wonder in language and in a taxonomic system. The 'incommensurable' objects lying in the museum are in constant tension with the science that tries to order them. Wonder lies in-between. By contrast, one could say that curiosity can be satisfied, that satisfaction is accounted for by a means of taxonomy and classification and that an existing schema accommodates those classificatory means. Thus the incommensurable colonial artifact becomes a variation on an existing European model. We see this at work in the Hellenic casting of aboriginal subjects in colonial paintings of the heroic nude 'savage', or in indigenous mothers rendered in the guise of Renaissance Madonna and child, or in the Romantic landscapes that are made of New Zealand by William Hodges.[20]

---

[18] Notably Greenblatt's interest is an entirely bookish one reflecting the Eurocentric obsessions of eighteenth century literature.

[19] This is not just a phenomenon of uncomprehending discoverers to slip between 'curiosity' as embodied by a physical artifact and 'curiosity' as a mental state. This slip is common to ancient and contemporary aestheticians alike. To counter such confusion Irene Winter has translated the Sumerian descriptions of wonder instead as *ad+miration*, to account within the term for the 'visual *spectacle* and the *spectator's response*'. Cf. Winter 2000: 22–44.

[20] See Hodges 2004. An exhibition catalogue. The National Museum in Canberra makes the point in its display by contrasting drawings of classical sculptures with those of Aboriginal bodies.

In a state of wonder, as Wittgenstein writes, the subject is not, as it usually is, amidst the other parts of the world, but sees them *with* time and space. R. W. Hepburn in his commentary of Wittgenstein writes:

> [A]ny hint of being at ontological odds with that spatial-temporal object-world, of being incommensurable with it, may prompt us towards interpreting a field of experience as a 'world in itself'. Accepting such promptings, taking this subject-matter as a world, we are taking it as a proper object of existential wonder. Cosmos-wonder is transferred with ease to any strangely inassimilable micro-cosmos. (Hepburn 1984: 150)

I would argue that to separate objects from the world, to create microcosmic wonders is in fact the modern museum's hope for its artifacts. Wonder is something a museum display tries to contrive in the visitor's reception of images. There is a sense in the museum that these wonders remain to some extent other and unmastered despite the neutral scientific rigor of the museum's Linnean genealogy. In Hepburn's terms 'wonder doesn't see its objects possessively'. In contrast, 'curiosity-knowledge' is a kind of possession, 'a tick on the tourist's place-list' (Hepburn 1984: 134). Hepburn shows how our perception of the nature of the world relates to wonder. Precisely when we perceive nature and our otherness to it, the germ of wonder grows. We need only think of the Enlightenment conception of the sublime in landscape paintings by Caspar David Friedrich and his prodigies of German Romanticism such as Eugene von Guérard who then rendered the Australian landscape (1852–1882) in the mode of the sublime.

There was an unacknowledged inability to understand each other in the first colonial encounters in Australia that constituted the climactic moment of aporia. Without wanting to relegate wonder to the supralingual realm, its relation to a lack of understanding is compelling. The far more controlled pursuit of curiosity may drive an encounter where a mutual unintelligibility threatens scientific order. Yet as soon as the subject can wonder how the other understands, every foreign word and artifact can open out into wonder.

In the colonial context curiosity is aroused over the *use* of artifacts from other cultures. When we look at the *woomera*, the question of *how* it was used by Aborigines arises as another articulation in the museum of our curiosity. So what of the debased, grubby curiosity that was dragged victoriously from otherness and possessed? Does the 'novelty value' identified in curiosity by Kant hinder its philosophical investigation over the test of time? Can we best define curiosity negatively (by saying it is not wonder)? If we claim that there cannot be incommensurable curiosity, then we may say curiosity is accompanied by a desire for the closure that measure and classification offer. Where there is a lack of closure there is wonder.

Opening a discourse to new bricollage (collection and reassemblage), wonder is a force at the beginning of investigations. Seeing worlds as an outsider sees them, *sub specie aeternitatis*, and moving always to find a better place at the periphery, lack of closure means also a lack of consistency in the positions of the wandering, the wondering, and the hybrid.

The displacement between the spaces in which a piece of material culture 'lives' has the potential to give an artifact new life beyond the death incurred by its extraction from a living culture.[21] Thereby the spaces in-between, such as museums and their collections, gain an incommensurability of origin that is wondrous. This wonder focuses consciousness on a liminal (or in-between state), giving it agency in recombining spaces and temporalities for contemporary hybridity. Homi Bhabha speaks of the 'temporal breakup' that is inscribed in these 'in-between' spaces. Within the 'new international space of discontinuous historical realities' he says we are liminal (Bhabha 1994: 310).[22] This being in-between may function *differently* in irreconcilable spatio-temporal structures. In-between the Aboriginal Dreamtime (in which there is no linear 'time') and the colonial sphere in which our chronological practices have likewise determined

---

[21] Though in the context of the museums founded on French colonialism Chris Marker's and Alain Resnais' film *Even Statues Die (Les statues meurent aussi)* of 1950 (and censored for more than 10 years) beautifully portrays the fate of an art that was once integral to communal life but became debased as it fell victim to the demands of another culture.

[22] The 'signification', Bhabha writes (1994: 310), of 'interstitial passages and processes of cultural difference' 'must be mapped'. This liminal space is open to potential, or what Nietzsche termed 'return', Heidegger described as 'thrownness', or Derrida found to be a 'trace'. If adopted into Bhabha's treatment of culture as a site of interrogation, the epistemology of colonial objects in the future could be in the hands of the wondering interpreter to find openness to the potential that at any time it may also open elsewhere.

Western practice, Australians may find new agency.[23] The atemporal indigenous cosmology absorbed the first sighting of the *Endeavour* as if with an intuition that wonder is directed at what can be pursued endlessly. Being in sacred time is being without time, and yet an individual can be part of both systems, can slip into no-time, or count themselves part of the arbitrary measure of hours. Consciousness formed in-between temporalities is a way the seemingly insurmountable border between non-Aboriginal and Aboriginal cultures could be crossed, and crossed out.

Post-structuralism's attempt to break down binaries may find new ground where 'in traditional Aboriginal thought there is no nature without culture, just as there is no contrast either of a domesticated landscape with wilderness, or of an interior scene with an expansive "outside" beyond four walls' (Bennett 2006: 518).[24] There is vitality of deconstruction in Gordon Bennett's painting and writing (of which this is an example) that is a violent contrast to the collection practices of previous centuries. It makes rigid classification seem like a desperate symptom of the epistemic regimes that govern the museum.

Locating culture as Bhabha does in a negotiation of the incommensurable is especially liberating to those in Australia whose agency has been thwarted by the liberal representation of Aboriginal

---

[23] For a compelling treatment of the Marxist view that changes in technology determine culture and as that relates to language and creolization in colonial Africa, cf. Glissant 1997.

[24] Within Australia arguably the landscape is a shared discursive space – it is a referent available to all Australians and made sensible in hybrid artistic forms. Richard White (1981) has argued this at length. The notion of a shared or hybrid discursive space is a way in which this Australian material differs from Dean and Liebsohn's critique of hybridity: cf. Dean and Liebsohn 2003. Dean and Leibsohn argue that what is at stake in the notion of hybridity is 'our willingness (or need) to "see" the influence of European culture', when hybridity may be largely 'invisible'. In other words the recognition of indigeneity in art is driven by our desire to see that indigenous people were not completely vanquished. The term hybridity has the political charge of a discourse that deals with the power relations that engendered the conditions of encounter. The transformations that have occurred in indigenous art since first contact are therefore not recognized because they are not visibly 'indigenous'. This is certainly the case in the contemporary Australian art, especially by 'urban Aboriginals' who are not isolated from 'modern' media. Australian artists such as Julie Gough break with the stereotype of what traditional Australian art looks like – dot paintings on the body and bark, or since 1979 also on canvas.

culture as alien.[25] Politically convenient but unimaginative, this position has ossified attempts at fostering mutual understanding through 'reconciliation'. Now it begs to be asked within the highly politicized reconciliation process in Australia whether it is really necessary to 'understand' each other in order to have successful coexistence?

The unreconciled languages, trade relations, and understanding of art and ritual as instantiated in that first encounter resonate in Australia's continued failure to cross between these cultures, but it is not a justification for the future extinction of Aboriginal culture. As a strategy of survival hybridity need not conspire in the 'assimilation' of all Australian cultures.[26] It has been argued by Ian Anderson that the notion of cultural hybridity was a way for white history to equate the ambiguity of 'Aboriginality' with the notion of being 'without culture' (Anderson 1993–94: 10–12). Historically this may be true, however in our current zones of contact the location of culture now is the consciousness of those who move between interstitial spaces.[27]

The cultural predisposition of the West is toward anxiety if we do not share language and its way of structuring time and space. It is destabilizing but also wondrous to think that two cultures may be incommensurable and yet not incompatible. How might we redirect the anxieties about the terms of exchange across cultures that differ

---

[25] Julie Gough's summary of an email exchange we had about the liberating function of hybridity in art as it seeks to dismantle an essentializing impulse is published online in *Machine*, Issue 1.9, (http://www.artworkers.org) accessed on July 16, 2007: 10.

[26] Ellen M Smith has critiqued the relevance of Bhabha's theory of hybridity when in relation to Indigenous and not migrant culture – in response to Anderson 1993-94: 10–12.

[27] Here I am borrowing the title from the 2006 Biennale of Sydney 'Zones of Contact', director Charles Merewether stated that: 'the condition from which the work of art appears is specific and yet, contains the potential to cross boundaries, to be understood elsewhere and animate a sense of history as defined not simply by force of separation but by threads of entanglement. To experience the work of art is to enter an interval between time, a pause in time. In so doing, it offers a different ways of seeing who we are and a new zone of contact.' *2006 Biennale of Sydney,* http://www.bos2006.com/ (accessed on June 1, 2006). The Biennale's thrust was towards art by those criss-crossing spaces of dislocation and displacement and by those living in another's culture. The works explored the influence of different cultures upon each other, as well as the land shared. From zones of conflict around the world the international artists reflected how negotiations of fraught territory resonate with the history of land and culture in Australia.

both materially and immaterially in their understanding of the world?[28] Bhabha's and Robert Glisson's models for language could inform a map of both the Aboriginal and non-Aboriginal Australian subject and how it can relate to the material culture of the land. Gordon Bennett expresses Glisson's notion of 'errantry' when he writes that 'the path to "no place" [utopia] is one of attitude, a sensitivity to nuance and the possibilities of evolution toward a kind of ecological awareness in thought, and therefore in action. However, it is the path that is the key and not any final destination' (Bennett 2006: 519).

A struggle against the notion of an original is also paramount in acknowledging the mobility of objects. Though different from a human in its relationship to language, an object can be seen as errant, as a nomadic in-between, because of its crossing of tangible and intangible borders. Bhabha's model of cultural liminality in the 'new international space of discontinuous historical realities' means that an Aboriginal Australian can be one (id)entity in relation to the institution from which they make a living – say the art market in the case of a contemporary Aboriginal artist – and another in relation to their clan (Glissant 1997). Important is our rethinking of what has historically been seen as an impossible combination between the indigenous and modern. At the emerging periphery of cultural production contemporary artists challenge the claim of essential identity. Their work displays inconsistencies both within and in moving between and beyond the binary terms of indigenous/modern, traditional/contemporary, artisanal/academic.[29] Yet these inconsistencies do not undermine but rather open them and their artwork to recombinant hybridity.

---

[28] The UNESCO convention on intangible heritage acknowledges that material and immaterial culture is not separable. In lieu of this 2003 convention (which is still in the process of being translated, existing at present in draft in 42 languages), a study of indigenous heritage was carried out by Erica-Irene Daes in 2000 (see Brown 2003: 225). Also see the upcoming publication of the conference on *Intangible Heritage*, Harvard University Faculty Club, 4–6 May 2006.

[29] Akhil Gupta problematizes the term 'indigineity' through the study of 'hybrid' uses of language and technology. Gupta points out that 'indigeneity' is defined by negation – it is by definition 'not modern' but bounded and local in the old dichotomy between traditional and modern. Furthermore, 'indigeneity' has been 'fetishized' and the politics of calling someone indigenous is a risky strategy because of the kind of taxonomy it adheres to (cf. Gupta 1998).

The *woomera* I threw in at the beginning of this story therefore returns.[30] In Christian Bumbarra Thompson's *Emotional Striptease* (Figure 3) a *woomera* is cradled by a young indigenous woman and the terms of 'incommensurabilty', 'curiosity' and 'recombinant hybridity' are revitalized in artistic practice. The temporalities of this narrative are also interwoven in these photographs. The backdrops to *Emotional Striptease* are the fragmented citations of Australian landscape in Melbourne's postmodern architecture. The institutions are flattened as backdrops – The Melbourne Museum, Australian Center for Contemporary Art, Federation Square – and before the buildings are indigenous people that deploy their own urbanity. By commanding our attention with the *re*presentation of the objects and themselves, the striptease intervenes in the identity politics of museum display as a kind of institutional critique. So unlike the popular image of indigenous people in ethnographic photography in this display of agency, they all hold an artifact of their material culture from the Melbourne Museum's collection. It is the way they bear this artifact, though, that makes it an emotional striptease. The gaze is no longer

[30] Sartre wrote that the violence of colonialism returns like a boomerang. In his 1961 prologue to Fanon's *Wretched of the Earth* he says, 'it is the moment of the boomerang; it is the third phase of violence, it comes back on us, it strikes us, and we do not realize any more than we did the other times that it's we who have launched it'. The boomerang was already a metaphor for the movement of time in Ralph Ellison's *Invisible Man* (1947): 'My hole is warm and full of light. Yes, full of light. I doubt if there is a brighter spot in all New York than this hole of mine, and I do not exclude Broadway. Or the Empire State Building on a photographer's dream night. But that is taking advantage of you. Those two spots are among the darkest of our whole civilization – pardon me, our whole culture (an important distinction, I've heard) – which may sound like a hoax, or a contradiction, but that (by contradiction, I mean) is how the world moves: Not like an arrow, but a boomerang. (Beware of those who speak of the spiral of history; they are preparing a boomerang. Keep a steel helmet handy.) I know; I have been boomeranged across my head so much that I now can see the darkness of lightness. And I love light. Perhaps you'll think it strange that an invisible man should need light, desire light, love light. But maybe it is exactly because I am invisible. Light confirms my reality, gives birth to my form. A beautiful girl once told me of a recurring nightmare in which she lay in the center of a large dark room and felt her face expand until it filled the whole room, becoming a formless mass while her eyes ran in bilious jelly up the chimney. And so it is with me. Without light I am not only invisible, but formless as well; and to be unaware of one's form is to live a death. I myself, after existing some twenty years, did not become alive until I discovered my invisibility.'

averted from the exchange. Their performance reclaims the dead from the museum. In their Victorian period costumes they ridicule the contorted superiority of the settlers and wear the uniform of power and civilization. This recombinant appropriation of temporalities shows how even a spearthrower may help us cover more ground when we wonder about colonial *curiosities*.

## *References*


Anderson, Ian. 'Re-claiming Tru-ger-nan-ner.' *Art Monthly Australia* 66 (Feb–Dec 1993–94): 10–12.

Bennett, Gordon, 'Aesthetics and Iconography: an Artists Approach.' *The Anthropology of Art*. H. Morphy and M. Perkins (eds). Malden: Blackwell 2006. 518–19.

Bhabha, Homi K. *The Location of Culture*. London: Routledge, 1994.

Brown, Michael F. *Who Owns Native Culture?* Cambridge: Harvard University Press, 2003.

Clendinnen, Inga. *Dancing with Strangers*. Melbourne: Text Publishing, 2003.

Dean, Carolyn, and Dana Liebsohn. 'Hybridity and its Discontents.' *Colonial Latin American Review* 12/1 (2003): 5–35.

Deleuze, Gilles. *Différence et repetition*. Paris, Presses Universitaires de France, 1968.

Douglas, Bronwen, 'Art as Ethno-historical Text: Science, Representation and Indigenous Presence in Eighteenth and Nineteenth Century Oceanic Voyage Literature.' *Double Vision*. N. Thomas and D. Losche (eds). Cambridge: Cambridge University Press, 1999. 65–99.

Foigny, Gabriel de. *The Southern Land, Known*. Transl. David Fausett. Syracuse: Syracuse University Press, 1993.

Glissant, Edouard. *Poetics of Relation*. Transl. Betsy Wing. Ann Arbor: University of Michigan Press, 1997.

Gough, Julie. *Machine* Issue 1.9 (http://www.artworkers.org) accessed on July 16, 2007.

Greenblatt, Stephen. *Marvellous Possessions: The Wonder of the New World*. Chicago: University of Chicago Press, 1991.

Gupta, Akhil. *Postcolonial Developments: Agriculture in the Making of Modern India*. London: Duke University Press, 1998.

Hepburn, R.W. *'Wonder' and Other Essays: Eight Studies in Aesthetics and Neighboring Fields*. Edinburgh: University Press, 1984.

Hodges, William. *1744–1797: The Art of Exploration*. New Haven: Yale University Press, 2004.

Kant, Immanuel. *Kritik der Urteilskraft*. Wiesbaden: Suhrkamp, 1957.

Marx, Karl, and Jean-Baptiste Ganne. *Das illustrierte Kapital*. Salzburg: Fotohof Edition, 2003.

Mauss, Marcel. *The Gift: Forms and Functions of Exchange in Archaic Societies*. Transl. Ian Cunnison. New York: Norton, 1967.

Montaigne. *Complete Essays*. Transl. Donald M. Frame. Stanford: Stanford University Press, 1958.

Mulvaney, D. J. *Encounters in Place: Outsiders and Aboriginal Australians 1606–1985*. St Lucia: University of Queensland Press, 1989.

Myers, Fred. *Painting Culture. The Making of an Aboriginal High Art*. London: Duke University Press, 2002.

Nordyke, Eleanor C. *Pacific Images: Views from Captain Cook's Third Voyage*. Honolulu: Hawaiian Historical Society, 1999.

Thomas, Nicholas. *Cook: The Extraordinary Voyages of Captain Cook*. New York: Walker and Co., 2003.

—. *Entangled Objects: Exchange, Material Culture, and Colonialism in the Pacific*. Cambridge: Harvard University Press, 1991.

White, Richard. *Inventing Australia: Images and Identity 1688–1980*. Sydney: Allen & Unwin, 1981.

Winter, Irene J. 'The Eyes Have It.' *Visuality before and beyond the Renaissance*. Robert Nelson (ed.). Cambridge: Cambridge University Press, 2000. 22–44.

Wittgenstein, Ludwig. *Notebooks, 1914–1916*. (Anscombe transl.) Chicago: University of Chicago Press, 1961.

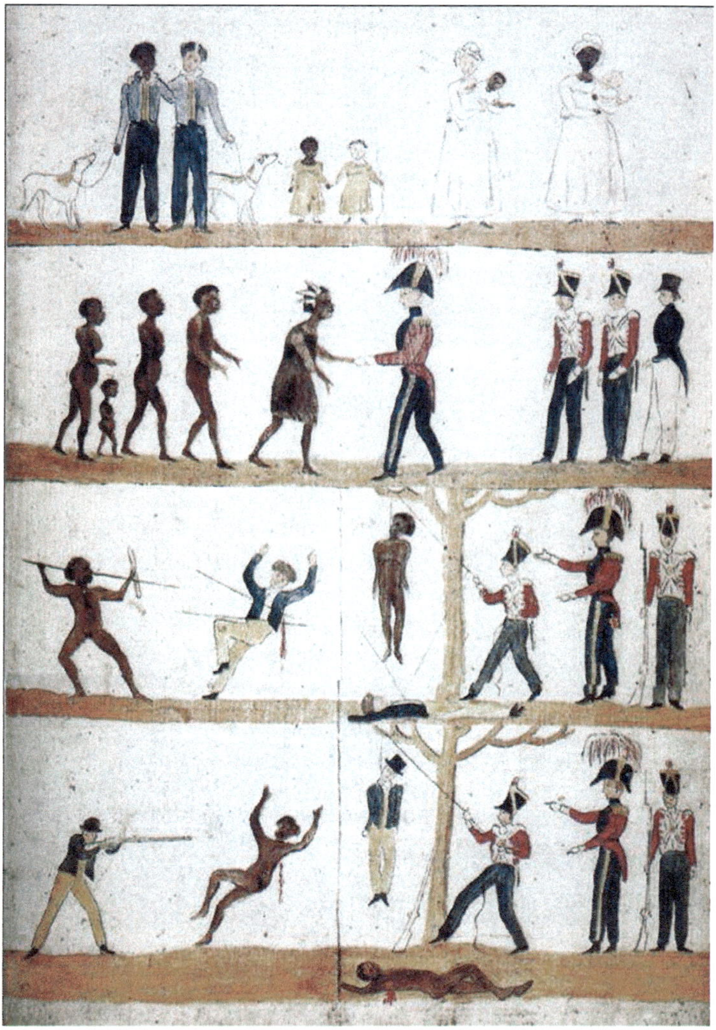

Figure 2: Unknown, 1830, *Governor Arthur's Proclamation to the Aborigines*. Wood panel nailed to trees in Tasmania during the Black War, according to the label on the back 'to communicate, not knowing their language'.
Peabody Museum, Harvard University.

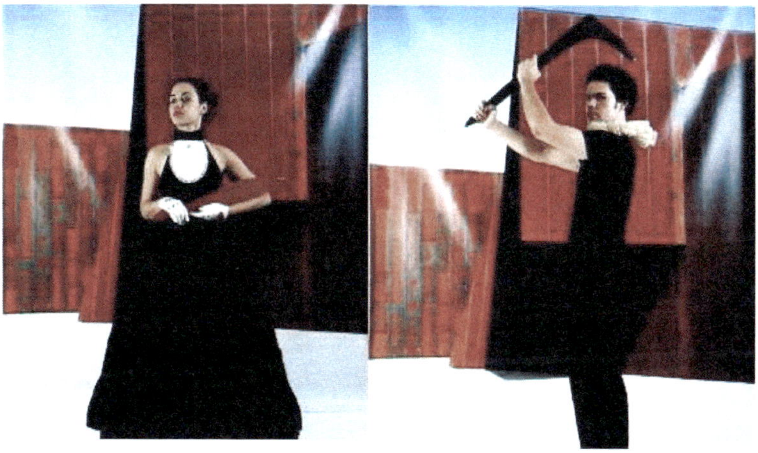

Figure 3: Christian Bumbarra Thompson, *Emotional Striptease*
Photographic series, 2005, courtesy of the artist.

# To Think of Myself as Statistic: Migration, Selfhood and the Australian Nation

Uli Krahn

*Using tools*

Recent academic discussions of space have focussed on the social ordering and manipulation of space which creates a material reality of place, as well as the transformation of space itself by means of technology (Casey 1988). Robert Sack defines place as: '[A]n area of space that we bound and to some degree control with rules about what can and cannot take place. Place can be any size from the small-scale of a room or a sacred grove, to the larger scale of a farm or city […] some of the smaller ones can even move, as in the case of a railroad car' (Sack 2003: 4–5). He considers the construction of place a universal and 'essential tool' that, as tool, he argues to be 'as pervasive as language' (Sack 2003: 4–5). I've started with Sack's account, as he concerns himself with space in relation to morality. Accordingly, he argues that, while acknowledging the various influences of embodiment, cognition, social and economic conditions, some personal agency must be assumed for discussion in a moral context. There is a philosophical problem here. As Sack asserts (and many colleagues would agree), 'we are manipulating the environment […] to have reality become what we think it ought to be' (2003: 4–5). There would be little point rearranging the world around us, if we didn't assume that, in turn, this environment is an important influence on us. However, considered in traditional terms, this tension becomes a little like the question of whether the chicken predates the egg: how can we make our environment, if it shapes us? How and when do people make that jump which enables them to imagine things to be different and act accordingly?

This problem recurs, differently-shaped, in various academic fields, especially those most affected by postmodern doctrines of the social construction of reality. For instance, Foucault's model of a society shaped in all its aspects by the exertion of power, leaves a gap many have observed about how to explain resistance to power in any form, or even how to observe that state of affairs. In the mind-related sciences, similar debates oppose various forms of biological determinism to free will. Free will, as a Christian concept, is not fully congruent in meaning with agency, even though they are easily mistaken for one another (cf. Flanagan 2002: 99–159, Frank 1986). Usually, and across disciplines, there is a tendency for discussions of agency to be split into mootable dualisms; I hope to show that the academic deadlock which often ensues might be avoided if questions were asked from a different angle. Surveying the literature in fields as diverse as cognitive neuroscience, philosophy and the humanities, you detect an undertone of nostalgic regret for a form of agency which, presumably, modernity has banished into the past, together with Christian faith, an enchanted nature and the possibility and ideal of self-knowledge. This undertone is a belief, not researched science, an emotional paradigm, if you will.

This postmodern emotion stands somewhere between the apocalyptic, the arrogant and the helpless. Viewed from a distance, debates evidently operate within a narrow Judeo-Christian metaphysical framework, albeit often with reversed tenets. Julian Bell describes this modern/postmodern certainty by means of uncertainty in the artist: '[T]to be detached (even to be despised) might be a guarantee that you alone possessed a new, unprecedented level of engagement with reality' (Bell 1999: 112). We are meant to stand apart from history, incomparable both in our knowledge of the world, and our experience of suffering. Susan Neiman contracts this neatly in her discussion of evil:

> If enlightenment is the courage to think for oneself, it's also the courage to assume responsibility for the world into which one is thrown. Radically separating what earlier ages called natural from moral evils was thus part of the meaning of modernity. If Auschwitz can be said to mark its ending, it is for the way it marks our terror. Modern conceptions of evil were developed in the attempt to stop blaming God for the state of the world, and to take responsibility for it on our own. The more responsibility for evil was left to the human, the less worthy the species seemed to take it on. We are left without direction. Returning to intellectual tutelage isn't an option for many, but hopes for growing up now seem void. (Neiman 2003: 4)

The displaced person, migrant, refugee, or whatever else governments decide to call foreigners, appears as one of the major symptoms of this perceived crisis of (post)modernity. Recent historical research has condemned this conviction as myth – apparently migration, far or near, brief or permanent, has nothing to do with crisis at all. It happens everywhere and all the time and ensures economic wellbeing (Lucassen and Lucassen 1997: 9). But it also seems to always create problems. Individuals or groups changing place threaten both the community that is left and the community entered. This is not surprising. Modernity's invention of the nation-state posits the nation as an organic bond of people and place, combined with an ideally far-reaching history. This ideal cannot explain either why somebody would wish to leave, nor accept the harmonious co-presence of different peoples within the one nation. It must, however, be noted that the nineteenth century nation-state was based on the operation of two opposed conceptions of space; the profoundly relative 'unique' space of the home soil was kept apart from the homogenous and absolute space of the booming natural sciences, which was keenly applied to places belonging to other people.

This tension within the definition of nation-space became particularly noticeable in 'homes away from home', settler colonies like Australia. As the home soil had been left behind, notions of race had to be more vigorously enforced, to carry the extra weight of defining who was Australian. This is one of many reasons we are only beginning to discover a more complete history of Australia, its exploration, colonization and subsequent settlement by migration. As initial British settlement occurred during intense international competition for trade and expansion, the region's long pre-history in seafaring trade and intellectual speculation was erased in favour of a neat temporal beginning as artificial as the concept of *terra nullius.* This took ongoing effort; the ancient Aboriginal overseas trade networks, reaching as far as China, were violently restricted deep into the last century, until eradicated. Australian history was, and often still is, written as the tale of an English colonial enterprise alone. Intellectually, this also negates the baggage of ideas and preconceptions which had encrusted maritime exploration and antipodean fantasies and affected travellers' perceptions (cf. Foulke 2002: 68). Even before Dutch and Indonesian traders, European writers and scientists from Ancient Greece onwards had comprehensively defined what Europeans would find in Australia. The southern continent was where the normal order of things would be reversed, thus providing a speculative platform for imagining total difference. Early Greek geographers required antipodes to balance the northern continents, and worried about the effects of gravity; all kinds of accidental and amazing adventures supposedly led people to Australia; mediaeval scholars debated whether God would allow antipodeans to exist, if they had no chance to hear Christ's message of salvation.[1] Dante had his protagonists cross Hell to arrive in the blinding light of the Antipodes, beholding the many stages of Mount Purgatory, topped with Earthly Paradise. The place of which theoretical geography demanded that everything was opposite to its normal position, naturally suggested itself as a location where middling sinners could work for redemption, actively changing their soul's destination in the location furthest from Earthly Jerusalem: '[T]he little bark / of my poetic powers hoists its sails, / and leaves behind that cruelest of the seas. And I shall sing about that second realm / where man's soul goes to purify itself / and become worthy to ascend to Heaven' (Dante 1985: Canto I, ll. 1–6).

It can be hard to pin down the jumps and transformation of an idea through intellectual history. Once you are acquainted with antipodean myths, rich in content, structure and imaginative appeal – to turn things on their head can be a mode of thinking, imagination or even social protest (cf. Babcock 1978, Hill 1972) – the form of the subsequent history of 'discovery' and the construction of colonised Australia fits almost too neatly into the pre-existing imaginative frame. This may be the nature of exploration: 'Many historians recognize that discovery is not an instantaneous result of physical encounters so much as a series of adjustments between preconceptions and direct experience that does not match them' (Foulke 2002: 68).

The historian Valerie Flint found, after detailed readings of Columbus and other early explorers' journals, that a certain amount of preconception and imaginings might be necessary to make the unknown perceivable at all (Foulke 2002: 70). The texture of a new place can appear like an undifferentiated wasteland; falsely imagining the shape of church spires or any familiar object, real or fancied, cracks that shell of the new, and first distinctions can be made (cf.

[1] For more details, see Uli Krahn 2002: 78–90.

Campbell 1988: 228). We are familiar with this effect, when we orient ourselves on a bushwalk by memorizing the tree that looks like a cross and the rock that resembles a rabbit. According to art theorist Ernst Gombrich, this is almost inevitable: 'If there is one psychological disposition about which one can afford to be dogmatic it is our readiness to see faces in any configuration which remotely suggests the presence of eyes and corresponding features' (Gombrich 2002: 264). He also suggests a 'tendency towards zoomorphic interpretation of abstract shapes' (2002: 271). Perception, as well as exploration, seems to involve a dash of fantasy.

While seeing things that aren't there, helping us to see the things that are, may be less surprising in the context of exploration, the subsequent question about the validity of perceptions is a philosophical chestnut. If you ask, what then is true and false, you return to a version of the philosophical deadlock outlined earlier. Wittgenstein's late writings concern themselves with overcoming this problem, insisting that all acts of perception occur as part of a situation. As the *Philosophical Investigations* argues, you never just look at something by itself, but as part of a social action, a looking in order *to*. Wittgenstein describes this – and other – social actions as following rules, as in a game. Meanwhile, Wittgenstein's concept of family likeness extends this contextual definition of vision into the temporal space of individual history. You perceive a likeness from the basis of what you have seen before.[2]

This problem of vision returns us to the difficulties of agency. Wittgenstein insists repeatedly that while his view of perception refuses absolute epistemological certainty, which he considers a mistaken concept and desire in itself, it does not negate the possibility of knowledge, or truth. Similarly, being part of a context does not necessarily rob the participants of self-determination; only an expectation of absolute, royal independence, cannot be sustained. A loss of this position is only upsetting if you wish to speak from the position of the philosopher-king. This understanding of philosophy, however, is exactly what has led many to describe it as an endeavour both politically incorrect and unsuited to our time. This is, however,

---

[2] See also Gombrich 2002: 288, who argues that the effects of art entail a strong temporal dimension; the narrow centre of vision moves over the composition, which, like music, achieves effects through expectation.

only one way of doing philosophy. I would argue that at this point in time, it's self-destructive to abandon the possibilities of a discourse that can reach across the borders of academic disciplines, nations and religions. The 'western' philosophic tradition is anything but; its history reaches from Africa far into Asia as well as the European heartland and its colonies. The contemporary challenge is not so much about truth, but the possibility of dialogue (Rorty 1980: 394). Wittgenstein's late philosophy might well offer a blue-print for a rational debate that aims for a flexible, situational and applicable truth. Translating this into applications, however, cannot but be somewhat messy or even transgressive; the forms of academic discourse themselves prescribe an attachment to the nineteenth century knowledge Wittgenstein attempts to overcome (cf. Fish 2005: 271–83, Wilson 2004: 3–49).

### *Changing the Space of Oneself*

> During the last decade it has become more than clear to historians working in the field of migration that this phenomenon has to be regarded as a normal and structural element of human societies throughout history. Generally, migration is no longer viewed as a sign of crisis, as a phenomenon exclusive to the industrial period, as an element of the 'modernization' transition, or as a typically Western occurrence. Nor do scholars stress only political factors in explaining large movements of people. Finally, the time when peasants in pre-industrial Europe were perceived as a stable, non-moving, and sedentary world, or when Handlin portrayed immigrants as rootless and desperate lies behind us.
>
> The new paradigm, presented by Frank Thistlewaite in an embryonic format 35 years ago, teaches us that migration is part of the general human pattern, essential for the functioning of families and crucial to the operation of the labour market. (Lucassen and Lucassen 1997: 9)

The authors consider the main problems of migration history to lie in its dispersion across disciplines, and the difficulty of getting these to talk to one another. Furthermore, this 'widespread breakdown into myriad sub-specialties' encourages the creation of typologies which quickly settle into 'fixed dichotomies' which may cause 'the dividing and isolating capacity of an analytical framework to overshadow its clarifying and explanatory potential' (Lucassen and Lucassen 1997: 10). For instance, Lucassen and Lucassen consider it near impossible to distinguish free and non-free labour, and propose 'a continuum' from chattel slavery via serfdom and debt peonage to free labour

(1997: 12). They find that studies of the settlement process 'tend to concentrate on the negative aspect', where long-term studies suggest that overall outcomes for migrants are positive (1997: 21–2); and diagnose a related danger of focussing on ethnicity to the exclusion of other affiliations such as gender, work, or neighbourhood. Lucassen and Lucassen suggest as major source of confusion, that '[a]lmost all historians have always regarded the national state as their point of departure. In this perspective it is understandable that migration has been treated primarily as a sign of crisis or weakness' (1997: 36).

Migration thus poses various challenges to the nation, and in reverse, to those studying it, highlighting, as Lucassen and Lucassen do, how the humanities' focus on nation debilitated our knowledge in this case. For instance, many would be surprised to hear that in the nineteenth and twentieth centuries, France and the United States were on par as immigration destinations (Green 1997: 57–72); writing the history of the ancient *grande nation* and that of the 'land of the free' required very different perspectives on very similar social processes. Similarly, despite extensive available documentary material, the debate continues – unresolved as ever – as to who exactly the Australian convicts *were*, hardened criminals or victims of class and poverty. Even less is known about people arriving in Australia independent of British Government initiatives, including even rough estimations of number (cf. Schlomowitz 1997, Richards 1997).

Most relevant for my purposes, it is still 'largely unexplored […] how, from a pool of potential migrants, male or female, the particular persons who actually leave select themselves' (Hoerder 1996: 218), thus returning to the theoretical problem of agency. There exist broad statistical predictions of who is most likely to migrate, but there is conceptual and perhaps emotional difficulty when it comes to jumping from the level of statistics to the level of the individual. This, to my mind, is immediately related to monadic and individualistic definitions of selfhood and nation (Wilson 2004: 3–49), which make large areas of interaction invisible, or even transgressive, as they interfere with neat separations into known categories. The problem may not stop here, but dovetails into a fundamental problem of how to relate whole and part. This is something our current academic methods find difficult to deal with, and often a mysterious gap or lack – between self, other and world – needs to be invented.

Put in simpler terms, how can I think of myself as a statistic? (Beck 1986) The difficult thing here, both for my writing and the reader's understanding, lies in how to think in such terms; if inside and outside, body and mind, self and society, only exist in interaction, attempts to follow this knowledge from abstraction into detail requires some uncomfortable sliding around of traditional categories. You need to make your own line of sense here. In migration studies, such attempts are made, for example, in Donna Gabaccia's linking of Italian and Chinese migration patterns and experiences (Gabaccia 1997). The line of enquiry I am following looks at migration under the aspect of trauma; not trauma induced by migration, but migration as a means of breaking out of the confining cycles of trauma.

## *Changing Perspectives*

> [T]his is yet another juncture of thought at which a misconstrual of the self, as a kind of grammatical-optical illusion, makes us inquire into the nature of consciousness *itself*, i.e., in isolation from the human practices, engagements, and interactions that assure the intelligibility of the concept of consciousness in the first place, and this in turn causes us to deeply miscast the nature of autobiography [...]. (Hagberg 2004: 238)

How are we to imagine a self as created in context as Wittgenstein suggests? How is it to imagine itself, define or claim its consciousness? Context goes far beyond the social; it occurs on any level of observation, from molecular processes in the body, to acting according to the statistical patterns visible in large societies (Ball 2004). Body and society can both be described and predicted according to physical laws. (This is not the same as the full control or absolute understanding which some definitions of knowledge demand.) Discussions of the social, even from materialist traditions, tend to somewhat ignore this overlap, despite the exhilarating possibilities a shared language and platform of research could offer. The laws of statistics, like fractal shapes, operate reliably across micro and macro levels of observation and complexity. We, however, cannot make full use of this, as both the monadic traditions of academic discourse, and our concept of civil liberties, require – or seem to require – an independent, free-willed self. Furthermore, if knowledge derives from context, this seems to threaten the possibility of objective truth. David Stern argues that this is a misreading of Wittgenstein, that such charges against his conception of rule-following within forms of life 'operate with a misconception of objectivity that makes these observations about human agreement seem so threatening' (Stern 1995: 103). Following Wittgenstein's repeated exhortations that 'nothing is hidden', the problem of 'statistical selfhood' combined with consciousness and agency may not be as mysterious or seemingly impossible as theoretical debates would suggest; after all, we know the statistics, as well as examples of humans acting in unpredictable ways. The problem may be to find words for what is before our eyes.

> [T]he genuine understanding of another person's thinking, another person's thought, would be a metaphysical impossibility; in accordance with a far less neat reality, such understandings, sometimes characterized as genuine in contexts where that word marks an important contrast, will come in a thousand different forms (it is perhaps literature that best provides the vast catalogue of cases of other- and self-understanding of precisely the kind Wittgenstein repeatedly suggested we assemble) – reminders of what we *actually*, contra the picture, say and do – in order to change our way of seeing, to loosen the grip of the falsely unifying picture. (Hagberg 2004: 239)

Writing can play this role because it is moved along by the tensions of self and other, self-perception and being perceived by others, and identification, i.e., abandonment of self for a limited time. A text about myself gains a different truth when it is read; and while the writing self may have no privileged access to self-knowledge, the process of writing and reading intersperses a set of well-known rules in the tricky space between self and self-perception.[3] Richard Shusterman argues from a reading of Wittgenstein's writings on voluntary physical movement that the difficulties of bodily self-perception can be ameliorated by training and physical discipline (Schusterman 2003: 207). This may not be different for the less well-defined body parts that compose a text. Any form of discipline potentially provides a set of known rules which tells you what you are doing and how, just like a measuring tool.

## *Voices*

> *I cannot remember when I first heard the voices, they must have been there before my birth. I didn't mind them so much, they were company, like the*

[3] Cf. Iser 1993: esp. Introduction and Chapter 1.

*books and the toys and my brother. As I got older, my own voice got lost, and the other voices drifted further from me. Sometimes, they'd enforce hours of silence, especially when I was supposed to speak. I knew you couldn't have no thoughts in your head, but no matter how much it angered other people, they'd empty me out, like a camera. I didn't enjoy that, it hurt. At other times, they'd repeat the scolding I'd just received and add more to it. There was nothing I could do. Thoughts, moods, words, ideas would come to me at random or disappear. I didn't have any control. The same thing would happen to my body, which would change shape uncontrollably, as far as I was concerned.*

The above describes an unfamiliar and obviously pathological state of mind. On the other hand, most people have probably felt a little like that at some time in their lives. Perhaps the elements of self can drift too far apart to be controlled, temporarily or permanently. Far apart in this instance is not directly spatial, but more like the ability to move in body, which knows varying degrees between strength and helplessness.

*We lived in several places. One I didn't like and feared. A dark and faceless person woke me, sweating from nightmares, most nights there. He lived in the low sky, the muddy forest, the silhouette of the streets, the thick local dialect we spoke. It was good Mum believed in frozen foods. The proper and traditional food of our place made me ill with tasting of him. The other place made me happy, just being there, or thinking about being there, was as if I'd never known anything bad. There was no trace of him in the red food or the hills or the sea, so that a shabby and poor and not particularly pretty stretch of the Adriatic coast was my earthly paradise. It's strange to become a foreigner among your own people, but fortunately, it is possible. I probably couldn't have thought it up or survived if I hadn't known the other place.*

*I couldn't understand what was happening to me. While school and adults expected me to have increasingly more control over my actions, it was near-impossible to influence what went on in my head. Something had to be done. There were rituals to help; if I wore the same clothes I'd worn passing a test, I'd pass again this time. If I did that too much, though, it could spin out of control. I mightn't be able to change my clothes for weeks for fear of what bad might happen. It felt stupid and irrational to base my sense of who I was on props such as who I'd been a year ago in the same season; it would go okay for a while, but once I'd think about it, I'd remember that I'd done the same the year before, too, and the infinite regress felt like falling. It was hard to talk about these things, or even name them. I just wanted enough order to get through the day without too much pain. I could invent orders, but they'd never keep their shape long. Something was tearing at me, like a strong wind, and feasible modes of being, such as rituals, would turn into prisons, or explode if something bad happened, which meant I had to discard them and invent new ones. Fortunately, I could spend most of my time reading, the side*

> *effect being that school wasn't too much of a problem after all. Not only would I become the people in the book, the echo of the book's narrative voices would stay with me. With luck, the book's voice explained the world to me as well as it could; a lot depended on finding the right books. The book's voice would remain itself, but continue commenting and describing, and because I could think about the book from a critical distance, as well as entering it, I had an outside handle on what went on in my head. This made things a lot easier. It wasn't entirely safe, though. Some books could put you at the wrong angle to the world, and make things more dark and complicated. And there were situations that never occurred in books where it was hard to imagine what the book would say. Growing older, I worried what the books would think of me; they often didn't exactly celebrate teenagers living in provincial towns. Sometimes they couldn't help at all, because the force of the scolding and humiliating voices grew too strong. I didn't like being so dependent on the books, either, and tried other disciplines. Sport was tricky even though I liked to play, very much, the social dimension was too hard for me. I found a pleasant niche in the local athletics club; in a small town, individual high achievement is as frowned-upon as disability, and the kind people of the club created a space for both kinds to train together. Kind people were good, I could install them in my head, too, so they could guide me with their voices. A nice teacher might keep my overall frame of reference stable for a term, but the gravitational pull was always against me. Because I kept working at it, the constrictions remained flexible; sometimes the horror movie unfolded in my imagination, sometimes in my house and sometimes on the street outside. Shifting the fear around meant that I got to use each space at some time. I tried not to think about this, because part of it seemed normal, and part of it very wrong. There was no point trying to explain why I could get upset just looking at my handwriting and seeing from whom I'd snatched each letter and worrying what that meant.*

The most striking thing when I look over the above text is the ferociousness with which I continually scanned the environment for bits and pieces to use. I was very lucky that my abilities and environment provided methods that opened up opportunities instead of closing them. Almost, or perhaps an actual, biological need, like hunger, the need for inner structure can override considerations of personal safety. If instead of a decent library, my home town had offered only petrol sniffing for diversion, I would have done that for structure. From a clinical point of view, I'm a messy 'double diagnosis' of paranoid schizophrenia overlaid with post-traumatic stress disorder (PTSD), carrying an extremely small statistical chance of any positive outcome. It actually worked a little the other way, as the fantastic possibilities and detachment that came with the voices could be played off against the PTSD horror movie effects. I'm sharing these details to make it very clear that I am not writing the fiction to suit my theoretical convictions. The work went the other way, me continually trying to make sense of things and to survive and according to my propensities, approached it as an intellectual problem.[4] This, evidently, worked quite well, as I overcame enormous odds to be where I am now (if unfortunately unable to benefit from current anti-psychotic medication, which destroys your ability to reason).

*Statistical Self-Consciousness*

> [A] personal narrative [...] gives us a sense of self over time and in different situations. Psychological trauma occurs when the person loses this sense. Some things happen to us that occasion 'constructive bankruptcy' – the inability to plot the event in terms of our system of meaning. (Butt 2004: 113)

Trevor Butt's description seems to sum up the above autobiographical tale. On the other hand, as Butt himself points out, there is no objectivity to PTSD – it affects some people and not others who were exposed to the same objective trauma. Others suffer without having experienced anything life-threatening. According to Butt, trauma can only be understood 'in terms of how something appears to the person' (2004: 111). I would go further, and question whether this inability to form a coherent narrative of self is as profound and conceptual as Butt's description suggests. While, at times, I continue to experience an aching lack of self, as well as the need to keep making up new constructions, who I am remains remarkably stable. I may feel like I have nothing in common with yesterday's person, but I still enjoy the same foods and activities, dress the same and remain friends with the same people. Perhaps the sense of failure of self is more like an emotion than a structural change.

This doesn't mean to define selfhood as an epiphenomenon of another, unknown something. It does, however, mean to say that selfhood is possible, or rather, unavoidable, even where medical science declares it to be nearly impossible. The absence that hurts me seems more like a partial paralysis now, but paralysis of what? It's all bits and pieces, socially and spatially dispersed. On the other hand,

---

[4] For an interesting, and diametrically opposed outside view on psychosis and intellect, see Sass 2001: 98–155.

this should hold no surprises. It is precisely what various theories have been telling us for some time. The startling thing is that even when dispersal seems at its most chaotically complete, the one thing that will not stop is the relentless, searching agency of self.

William James suggests that, while the sense of self may be located in a specific location of the body (for which he received much criticism), the self itself is widely spread over space and time, encompassing your possessions, clothes, home, car, office, relatives, pets, friends and so on (James 1983: 1065–1107).[5] The idea that what we call self is a set of activities, possessions, experiences and relations spread through space is thus neither brand new nor, as in James' argument, needs to negate the possibility of self-determination or spiritual capacity, or a lack of commitment to rational method. Pursued to its consequences, though, it makes the distinction between inner and outer, self and world and self and other a matter of degrees, not absolute opposition. Perhaps the amount of agency is the sum of the varying degrees of freedom in each separate part; perhaps, following complexity theory, there are turning points where small quantitative differences cause large, overall shifts. The degree of dispersal in itself certainly does not seem to weaken or strengthen; highly traumatized people often seek a near-autistic minimal environment, and still experience the sufferings of self-insufficiency (Butt 2004: 113). The other way around, it is a cliché of our time, that the rich and powerful are extremely mobile, both spatially and socially, and experience this dispersal as pleasant.

Most notable here is that continuity or coherence may not be important for functioning or well-being at all. The emerging picture of the self – if we have to have one – might look more like a flock or herd of varying density and varying intensity of attachment, the two not necessarily overlapping one another; for instance, the time and energy you spend at work may pale in importance beside the four hours you spend twice a year on the stage of your amateur theatre, as far as self-definition is concerned. Interestingly enough, you could fit this well into recent theories of computational intelligence, which focus on self-organizing systems of intelligence that are modeled on animal flocking behaviour (cf. Bonabeau *et al.* 1999, Riegas and Vetter 1990). Similarly, as in the work of Niklas Luhmann, the auto-

---

[5] For a related contemporary view, see Clark 1997.

poetic systems approach to society, culture, economy, religion and selfhood is increasingly attracting interest in English-speaking countries, as it offers as yet under-explored possibilities for researchers wishing to escape old dichotomies of nature or nurture, body or mind and so on. By now, I have probably used enough imagery and metaphor to describe consciousness, agency and thought, to have fully disqualified myself in certain philosophical contexts; while I'm aware of the temptations that arise when speaking thus, I'll have to opt for practical application over theoretical purity.

*I could reduce the voices' intensity, almost turn them into happy babble, once I started feeding them foreign languages. Babel gets boring, so after a while I studied a single language intensely. It was a relief, thinking in English gave me several modes of control. If the voices scolded, I had arguments to shut them up, unless they spoke English. After all, I had to learn it for school. That reduced the nasty echoes in local dialect, which affected me most directly and intimately. When the voices would go on at me in English, it felt less immediate and even better, I could interrupt them when I caught them out on wrong grammar or vocabulary. I still had to keep working on new definitions of myself, all the time, as I kept falling apart from small impacts, or when the orderly shapes I'd made up turned against me. The shifting of languages was, however, enough, so that, voices and all, I could finish school and start university.*

*Being too weak of self begins as a bother and can reach a stage where it becomes an unbearable pain. Much of this has simple, practical reasons. When I'm bad, and in physical pain, I can lose all sense of a different past or future. This can turn even minor illnesses or stressful situations into big emotional crises, as I don't have the means to convince myself that things ever were or will be different. It also means that I am annoyingly sensitive to other people's emotions, so that social interaction becomes confusing, draining and sometimes impossible. Particularly when the other person isn't happy, but doesn't say so; only careful later analysis of the situation, often with a helpful person, enables me to avoid paranoid mal-conclusions. I am often extremely uncertain as to which is the right social rule to apply. It's easier in situations where the rules are clear. Mental health professionals can't help much, as they expect me to have a natural shape, which I would either recover, or lose forever. They could not tell me the rules. It's better with religious people, who can often give advice that responds to my actual questions.*

*I had to leave. Every brick, every sound and word had become overgrown with dark associations. I couldn't bear to speak my own language, and avoided it as much as possible. I was very happy to come to Australia; it was linguistic heaven. The more I learnt the language, the more freedom to use my own brain I gained. I relished the new and different landscapes and found it easy to feel emotionally connected to them. They weren't touched by fear yet, and never would be quite as badly. To my surprise, I also went through a standard-issue course of homesickness after arrival. The clear air and empty space could make me dizzy with the lack of my friends, but it passed. I only overcame my horror of the German language once I was firmly settled here, and owned an Australian passport.*

## *Impossible Vistas*

> Self-determined migration provided the modest but attractive prospect of escaping what were often paralyzing norms and customs, of finding jobs at comparatively higher wages, or facing fewer social barriers and wearing fewer badges of inferiority. Men did not have to doff their caps when asking for a job, and women in domestic service could wear elaborate bonnets, reserved back at home for middle- and upper-class women. (Hoerder 1996: 222, footnote omitted)

Dirk Hoerder's statement reads easily, and the observation seems almost self-evident, but it hides a deep scandal or transgression, one among many reasons why the history of migration is riddled with oversights, and ideology-driven constructions, such as Nancy L. Green diagnoses in her comparative study of immigration history in France and the US. In the case of Australia, I would argue that one of the main outcomes of individualistic and nation-centered approaches to history is not only an undue emphasis on British subjects, but also an enduring heritage of 'blaming the land'. Ancient antipodean clichés reappeared as artistic and scientific visions of Australian nature, thus hiding the extraordinary nature of Australian *society*. Whoever they were, precisely, according to nineteenth century genetics and criminology, early white Australians and their descendents should have continued to descend the social and physical ladder into an abyss of immorality and physical disabilities. Instead, as many surprised observers noted, they became, on the whole, tall, healthy and self-determining (cf. Schlomovitz 1997: 143–9). Similarly, as contemporary historians increasingly rediscover, colonial Australia was comprised of people of many different nations and races. For instance, despite official war historian C. E. W. Bean's assertion that 'the population of this vast unfilled land was as purely British as that of the two islands in the North Sea which had been the home of its fathers', about 1,000 Russians alone were members of the first Australian Imperial Force deployed in the First World War (Govor 2005: 2).

The nineteenth century nation-state had two components: a home territory, and a race, which by means of great historical effort and willingness to fudge stories, could be combined into a tale of the eternal belonging together and mutual shaping of a nation, its location and its people (cf. Young 1995). This put pressure on settler colonies, as here eternal roots to the soil could hardly be claimed, and those who had them, the Australian Aboriginals, were being murdered and dispossessed. Thus, race, and to some degree, class, had to bear the additional weight of spatial rootlessness, pushing everything extraordinary or abnormal back into Australian geography and nature. As the violent dispossession of the original inhabitants proceeded, an element of the uncanny added itself, reflecting both violence and its denial by the perpetrators. The antipodean myths of a world upside-down were handy, as to many European observers, Australian society really did seem like a turning upside-down of everything they 'knew' to be right about race, class and lifestyle. Shifting this 'wrongness' onto the natural environment was both in keeping with the ideals of nation, and conveniently naturalized a set of ideological positions, thus rendering them both invisible, and relatively permanent. To this day, concepts of culture frame interactions between different cultures as something exceptional; many models of culture assume a pure centre and a fringe of interaction, which may be celebrated for its hybridity. Inevitably, this understanding of culture tends to view foreigners as potentially threatening intrusions. I would argue that such a pure national culture never exists, but instead always entails ethnically diverse elements in interaction, or, at the very least, a small group defining itself against its neighbours. Foremost, culture is not an essence or entity that can do these things itself, but something *people* do; where among these actions are we to imagine a culturally pure core? A writer who's never read a foreign book, perhaps?

I would argue that Australian society still suffers the consequences of this version of history. The perceived 'abnormality' of Australian nature has often worked to distance Australians from an emotional or rational understanding of Australian ecological processes. Furthermore, it cuts a significant part of the population out of an involvement in the debate over Aboriginal land rights and reconciliation. In an extensive discussion about the issue of belonging, Peter Read voices this, unfortunately rather typical, opinion: 'I want to feel I belong here while respecting Aboriginality, neither appropriating it nor being absorbed by it. There may be millions to whom the issue is simply irrelevant to their lives. Non-Anglo-Celtic Australians, migrants or children of migrants, may well feel neither guilt nor responsibility, but dwell here in the belief that no racial or ethnic group has or should have a prior claim to the land' (Read 2000: 15).

This probably unintended extension of nineteenth century racialism and its half-blind approach to history, I dare predict, will render discussions about land rights divisive, and irresolvably complex. It's hard to right history if you haven't yet written it properly. This point of view is also flatly contradicted by the late Aboriginal activist and leader Mum Shirl, who asserts in her autobiography that Aboriginal land rights are the business of every Australian, whether they arrived on the First Fleet or hopped off a plane five minutes ago (Smith and Sykes 1981). It may be a blatant cultural difference between a self-described 'university educated, urban, middle-class and Anglo-Celtic' (Read 2000: 5) person and somebody like me, who grew up in post-Holocaust Germany. Unlike many commentators on Australian history, I don't find the idea of inheriting historical guilt impossible to think about and live with. While, of course I wish all violent histories had never occurred, I find this yearning for national moral historic purity more disturbing than the responsibilities of guilt. As long as we can't accept that guilt, it seems unlikely that we can repay our debts to Aboriginals today; and they, the people of the present, are the only Aboriginals whom we, today, might be able to avoid doing wrong to.

> I stepped off the wood and waited
> The ocean heaved inside, splashering;
> it never cared for me, although a pretty blue.
> Suckle yourself into the earth I told the feet
> but the left heel had broken,
> stupid cyclone,
> how can you then stand straight.
> 'Tis a stormy crossing of – something, heaven knows.
> But I notice, swaying,
> that even here cylinders are black, and crowds smell.
> Socks get dirty, pots burnt
> so it must be;
> even pastor's collar is brownish
> – is it oldest shipdirt or newest underneathglobe dirt?
> And will it come off – it must
> or there'll be no presbyter's cherrycake.
> Perhaps it's a different cake here anyway,
> a rich honour. (Krahn 2000: 116)

*References*


Babcock, Barbara (ed.). *The Reversible World: Symbolic Inversion in Art and Society*. Ithaca: Cornell University Press, 1978.

Ball, Philip. *Critical Mass: How One Thing Leads to Another*. London: Arrow Books, 2004.

Beck, Ulrich. *Risikogesellschaft: Auf dem Weg in eine Andere Moderne*. Frankfurt am Main: Suhrkamp, 1986.

Bell, Julian. *What is Painting?* London: Thames & Hudson, 1999.

Bonabeau, Eric, Marco Dorigo and Guy Theraulaz. *Swarm Intelligence: From Natural to Artificial Systems*. New York: Oxford University Press, 1999.

Butt, Trevor. *Understanding People*. Basingstoke: Palgrave MacMillan, 2004.

Campbell, Mary B. *The Witness and the Other World: Exotic European Travel Writing 400–1600*. Ithaca: Cornell University Press, 1988.

Casey, Edward S. *The Fate of Place*. Berkeley: University of California Press, 1998.

Clark, Andy. *Being There: Putting Brain, Body and World Together Again*. Cambridge, MIT Press, 1997.

Dante. *The Divine Comedy II: Purgatory*. Harmondsworth: Penguin, 1985.

Fish, Stanley. 'Take This Job and Do It…' *Critical Inquiry* 31 (Winter 2005): 271–283.

Flanagan, Owen. *The Problem of the Soul*. New York: Basic Books, 2002.

Foulke, Robert. *The Sea Voyage Narrative*. New York: Routledge, 2002.

Gabaccia, Donna. 'The 'Yellow Peril' and the 'Chinese of Europe.' In Lucassen and Lucassen, 177–196.

Gombrich, E.H. *The Sense of Order*. London: Phaidon, 2002.

Govor, Elena. *Russian Anzacs in Australian History*. Sydney: UNSW Press, 2005.

Green, Nancy L. 'The Comparative Method and Poststructural Structuralism.' In Lucassen and Lucassen, 57–72.

Hagberg, Garry L. 'Autobiographical Consciousness.' *The Literary Wittgenstein*. John Gibson and Wolfgang Huemer (eds). London: Routledge, 2004, 228–250.

Hill, Christopher. *The World Turned Upside Down: Radical Ideas During the English Revolution*. London: Temple Smith, 1972.

Hoerder, Dirk. 'Migration in the Atlantic Economies.' *European Migrants: Global and Local Perspectives*. Dirk Hoerder and Leslie Page Moch (eds). Dexter: Northeastern University Press, 1996.

Iser, Wolfgang. *The Fictive and the Imaginary: Charting Literary Anthropologies*. Baltimore: Johns Hopkins University Press, 1993.

James, William. *Principles of Psychology*. Cambridge: Harvard University Press, 1983.

Krahn, Uli. From: 'The Ernaeid: A Minor Local Epic in Verse', in *Hecate* 26/2 (2000): 116–118.

—. 'The Myth of the Antipodes as a Generative Force.' *Southerly* 62 /1 (2002): 78–90.

Frank, Manfred. *Die Unhintergehbarkeit von Individualität*. Frankfurt am Main: Surkamp, 1986.

Lucassen, Leo and Jan Lucassen (eds.). 'Introduction.' *Migration, Migration History, History: Old Paradigms and New Perspectives*. Berne: Peter Lang, 1997, 41–57.

Neiman, Susan. *Evil in Modern Thought*. Melbourne: Scribe, 2003.

Read, Peter. *Belonging: Australians, Place and Aboriginal Ownership*. Cambridge: Cambridge University Press, 2000.

Richards, Eric. 'Migration to Colonial Australia.' In Lucassen and Lucassen, 151–176.

Riegas, Volker and Christian Vetter (eds). *Zur Biologie der Kognition*. Frankfurt am Main: Surkamp, 1990.

Rorty, Richard. *Philosophy and the Mirror of Nature*. New Jersey: Princeton University Press, 1980.

Sack, Robert. *A Geographical Guide to the Real and the Good*. London: Routledge, 2003.

Sass, Louis. 'Deep Disquietudes: Reflections on Wittgenstein as Antiphilosopher.' *Wittgenstein: Biography & Philosophy*. James C Klagge (ed.). Cambridge: Cambridge University Press, 2001. 98–155.

Shlomowitz, Ralph. 'Coerced and Free Migration …' In Lucassen and Lucassen, 131–150.

Shusterman, Richard. 'Wittgenstein on Bodily Feelings.' *The Grammar of Politics: Wittgenstein and Political Philosophy*. Cressida J. Heyes (ed.). Ithaca: Cornell University Press, 2003. 202–220.

Smith, Shirley and Bobbi Sykes. *Mum Shirl: An Autobiography*. Melbourne: Heinemann, 1981.

Stern, David G. *Wittgenstein: Of Mind and Language*. New York: Oxford University Press, 1995.

Wilson, Robert A. *Boundaries of the Mind: The Individual in the Frangile Sciences*. New York: University of Cambridge Press, 2004. 3–49.

Young, Robert. *Colonial Desire: Hybridity in Theory, Culture, and Race*. London: Routledge, 1995.

# Eternal Recurrence: Art, Pain and Consciousness

Ann McCulloch

*'Eternal recurrence'*[1] *is the belief that one's life will repeat itself forever and that the higher human being, the truly free spirit, will be glad, will celebrate every repetition of suffering as well as happiness.*

Nietzsche's theory of eternal recurrence has received attention from many perspectives. In first coming across this theory one is immediately, almost involuntarily, forced into a response. It simultaneously invokes the kind of response that makes one aware that self-consciousness has entered into the equation and has simultaneously questioned the nature and understanding of consciousness. It is an awareness of what it means to 'be' in the fullest sense, and what it means to be conscious of one's ontology. Eternal recurrence is a doctrine that fights the compulsion to forget and is formed from the cognitive knowledge that forgetting structures the nature of what is known. There is, as Paul Ricoeur explains, an uncertainty regarding the essential nature of forgetting and it is this uncertainty that gives the search its unsettling character. Ricoeur characterizes this search as being driven by fears of things being forgotten 'temporarily or for good, without being able to decide, on the basis of the everyday experience of recollection, between two hypotheses concerning the origins of forgetting. Is it a definitive erasing of the traces of what was learned earlier, or is it a temporary obstacle – eventually surmountable – preventing their awakening?' (Ricoeur 2004: 27). Ricoeur in his analysis of different kinds of memory and different kinds of forgetting and erasure sees Freud's representation of the unconscious as 'helpful in breaching the impasse

[1] Also referred to in English as 'eternal return'. I have mainly used the term 'eternal recurrence' in this essay.

of the analytical colloquy' (2004: 445). His analysis of 'Remembering, Repeating and Working Through' and 'Mourning and Melancholia' focuses on blocked memory and recalls Freud's remark at the beginning of the first work:

> [T]he patient repeats instead of remembering. 'Instead of': repetition amounts to forgetting. And forgetting is itself termed a work to the extent that it is the work of the compulsion to repeat, which prevents the traumatic event from becoming conscious [...] the trauma remains even though it is inaccessible, unavailable. (2004: 445)

The allure of eternal recurrence stems from its power perhaps to rescue us from the past, from its pain, from its selected memories and from a past that we are helpless within, in that we cannot change it. Married to the psychological insight that in certain circumstances entire parts of a forgotten past can return, a belief in eternal recurrence might be able to rescue the person from this kind of forgetting and therefore from the anguish that compulsive repetition causes. Raymond Belliotto acknowledges that it is difficult to know whether the doctrine of eternal return is 'a cosmological doctrine, a hypothesis, a moral imperative, a psychological test, a reaffirmation of the death of god, or an attempt at secular redemption from the nihilistic moment' (Belliotto 1998: 78). Like Belliotto I consider that the 'psychological testing' that the doctrine evokes is most pertinent. It is elucidated when Nietzsche's doctrine is understood in the context of his work as a whole. To comprehend its metaphor requires an acceptance of Nietzschean perspectivism (that knowledge is provisional and truth-finding occurs from multi-perspectives); that metaphor itself for the poet, and I would add the philosopher, is not a rhetorical figure but a vicarious image with which he has replaced the concept; ('[t]he sphere of poetry', Nietzsche argues, 'does not lie outside the world, like some fantastical impossibility contrived in a poet's head; poetry aims to be the very opposite, the unvarnished expression of truth, and for this very reason it must cast off the deceitful finery of so-called reality of cultured man' [Nietzsche 1999: 41]); that Dionysian insight, shared with Zarathustra, of the unbearable insight of life requires the veil of illusion or the Apollonian form in order to be experienced at all and the experience of eternal recurrence comes with the functioning of the will to power. Art, as Simon Schama notes, 'begins with resistance to loss', and that art 'like memory, is never truly solid and seldom free of melancholy ambiguity, for it presupposes the elusiveness, if not the outright disappearance, of its subject' (Schama 2004: 9–10). Freud points out in 'Mourning and Melancholia' that the work entailed in working through blocked memories can only happen if the person experiences mourning and achieves a separation from the loss of love and hate. Art perhaps does not work without a 'will to power'; it perhaps accepts that the craving to 'nail down transient experience'; to construct an artifact of consciousness; to enact a mourning of loss, of intensities of love or hate, is doomed to failure yet its struggle against disappearance is a heroic one that leaves traces of its defiance.

What is this doctrine and why does it demand a personal response followed swiftly by a self-consciousness that seeks a metaphorical representation? Nietzsche writes of this doctrine first in *The Gay Science.* Nietzsche imagines a demon that reveals what he calls 'the greatest weight'. He writes:

> What if, some day or night, a demon were to steal after you into your loneliest hour and say to you: 'This life as you now live it and have lived it, you will have to live once more and innumerable times more; that there will be nothing new in it, but every pain and every joy and every thought and sigh and everything unutterably small or great in your life will have to return to you, all in the same succession and sequence [...].' Would you not throw yourself down and gnash your teeth and curse the demon who spoke thus? Or have you once experienced a tremendous moment when you would have answered him: 'You are a god and never have I heard anything more divine.'
> (Nietzsche 1974: 341)

How could this be so? The call for honesty, the call for re-evaluation, makes one self-conscious of where one stands in relation to the life lived. The eternal return was experienced by Nietzsche as the supreme thought, but also as the supreme feeling, an intensification of both intellect and affect. Nietzsche calls to the free spirits to love their fate, to accept all aspects of our living, to affirm the joy and the woe and not to wish it otherwise. 'Becoming innocent' is this process of affirmation, to will a life that accepts all of what is, to wish for things to be otherwise is to deny the world and oneself; to seek revenge or to experience resentiment is the choice of the herd; the 'free spirit' or the higher human type wants nothing in his/her life to be different.

> My formula for greatness in a human being is *amor fati*: that one wants nothing to be different, not forward, not backward, not in all eternity. Not

> merely bear what is necessary, still less conceal it – all idealism is mendaciousness in the face of what is necessary – but love it.
> (Nietzsche 1967: 9–10)

The doctrine of eternal recurrence stops you in your tracks; it demands of you a certain horrible recognition that there is pain that will not be obliterated; it makes one recognize this suffering and embrace it as one would happiness – to know and accept its essence and become strong in the face of it. The self-consciousness one experiences is one based on felt recognition which in turn becomes the 'faculty of the future or the function of the future, the function of the new'. Confronted in this fashion one can live with strength, in not attempting to obliterate a past of pain, to forget it or to wish it otherwise – one is in control, or is moving towards the overcoming of self. This self-consciousness, nevertheless, has a complication for those who have repressed unbearable pain and subsequently are lost to themselves.

Gilles Deleuze, when commenting on this self-consciousness, notes that it is not true that 'the only dead who return are those whom one has buried too quickly and too deeply, without paying them the necessary respects, and that remorse testifies less to an excess of memory than to a powerlessness or to a failure in the working through of a memory[?]' (Deleuze 1994: 15). Eternal recurrence stands in a giddy position when countered by those who are repressed and that the suffering or abusive acts they cannot remember are repeated in some form again and again. Elsewhere Deleuze is more specific. He writes: 'When the consciousness of knowledge or the working through of memory is missing, the knowledge in itself is only the repetition of its object: it is *played*, that is to say repeated, enacted instead of being known.' Repetition here appears as the unconscious of the free concept, of knowledge or of memory, the unconscious of representation. It fell to Freud to assign the natural reason for such a blockage: repression or resistance, which makes repetition itself a veritable 'constraint', a 'compulsion'. Here, then, is a third case of blockage, one which concerns, this time, the concepts of freedom. Here, too, from the standpoint of a certain Freudianism, we can discover the principle of an inverse relation between repetition and consciousness, repetition and remembering, repetition and recognition (the paradox of 'burials' or buried objects): the less one remembers, the less one is conscious of remembering one's past, the more one repeats it – remember and work through the memory in order not to repeat it' (Deleuze 1994: 15).

The kind of repetition evoked here is not that adhered to in the doctrine of eternal recurrence. This is not the acceptance of one's life but instead the denial of its pain. For Freud, the compulsion for repetition points towards the death wish. Paul-Laurent Assoun in his study of Nietzsche and Freud recognizes the difference. He writes: 'Whereas Nietzsche's repetition has a value of immediacy – that is why it resonates as authenticity and innocence – Freud's repetition presents a reflexive mechanical aspect, as an irrepressible character that one rejects and that returns with a tenacity that is troublesome more than painful, and which takes its whole meaning in the return of the repressed' (Assoun 2000: 160).

One must remember that Nietzsche wanted to be an affirmer; he repeatedly praised lightness of spirit, and wrote much about dancing and laughter. When Nietzsche put forward his strange theory of the eternal recurrence of all things – round and round again – this was significantly a rejection of gloomy nihilism and a way of saying 'yes' even to his physically painful and painfully lonely life. Yet he was not saying 'yes' to the kinds of repetition that comes with repression. He was demanding, instead, an honest confrontation with terrible pain. Sophocles' Oedipus is tragic and magnificent in his refusal to avoid murdering his father and marrying his mother. The self-imposed loss of his eyes when the facts unfold heralds a moment of nausea, the uncoiling of the black snake, but it also promises new courage and a future action that will, metaphorically, allow him to tear off that snake head and devour it in the fashion of Zarathustra.

It is often argued that Oedipus tears his eyes out because he cannot bear to see the truth. I would argue that the eyes have become redundant – that he is his own truth, he is now without fear. Freud notes in his essay 'The 'Uncanny' that man is beset from childhood with frightening things – with shadows that excite and arouse dread and horror. These may be things that are 'unfamiliar', strange or even uncertain. The 'uncanny' it seems also embraces the idea that there are certain things that 'ought to have remained secret and hidden but have come to light' (Freud 2001: 934). Freud in this essay analyses E. T. A. Hoffmann's story 'The Sandman' which among other things involves the threat that if children do not sleep the sandman will extract their eyes and feed them to his children. Freud's analysis of the story entails the argument that the fear of losing eyes is synonymous with the fear of being castrated by 'the father'. What interests me in this story is the attention given to the experience of repetition which enforces the feeling of 'uncanniness'. Freud refers to the way we all experience an inner 'compulsion to repeat' and that this experience is 'uncanny'. As such it 'ought to have remained a secret and hidden but has come to light'. Its coming in to the light is a moment of high intensity; it is a moment of eternal recurrence, when there is recognition and acceptance.

I am not concerned with the traditionalist view that tragedy usually enacting a death completes itself as an action with a moral order being returned to the immediate society that has been subverted. What interests more is that the tragic protagonist in his struggle to find his 'truth' is beset by attacks of unbearable insight as he/she struggles towards some understanding. For Nietzsche the insight was a Dionysian one, one so horrific, that art must disguise it in an Apollonian garb. The point is that the tragic character is beset by the need of strength and passionate recognition of all that he or she is.

The doctrine of eternal return is only understood, as Nietzsche remarked, when it is achieved in practice. Tragic drama tells the story of those who are led to the place where they are able to remember and able to accept their selves and their pasts – the unbearable insights of life – and then in that moment of cathartic experience acknowledge and act out the affirmation that dictates the doctrine of eternal recurrence. 'It is the power, the extreme intensity of this moment that transforms' (Lingis 1999: 60, paraphrasing Nietzsche in *The Will to Power*, 1058). The closest possible representation other than personal experience of such moments is that which is represented in art. At the beginning of this paper I indicated that the person most embodying the Freudian nature of repetition, that is, the need to repeat that which is not remembered, is the abused child.

A recent film by Gregg Araki, *Mysterious Skin,* is a work of art in its capacity to demonstrate this kind of repetition; it shows, also, the bleak life that the abused child enacts as he or she moves into adulthood. It also takes us at the end to 'the extreme intensity of this moment that transforms'. Like all works of art that confront a disbelieving public, or question civil and moral laws, there was a negative reaction from some sections of the public to this film. Indeed, it was even suggested that the film is less a critique of a society in which child abuse is rife than a kind of recipe, providing instructions to potential pedophiles about how to seduce young children. Gregg Araki's eighth film employs the art of the tragedian as he creates characters driven by their fate and yet simultaneously searching for how and when their free choice was invalidated. Araki creates filmic images and dialogue within a narrative that deals with the repression of memories that nevertheless are seen to feed the choices left open to people damaged by their experience. The film therefore is an artifact of consciousness dealing with the place and space of memory in consciousness itself.

Sexual abuse of children is part of our society but it is a subject area that is distasteful as well as a form of behaviour difficult to eradicate; it tends not to be treated as openly as it might be. Sexual predators of children are skilled at forcing children to keep the abuse a secret. What is evident, however, is that this is a secret that will do intense harm to the child, not only during the time of the abuse, but to the wellbeing of that child as he or she progresses into adult life. There are studies that argue that sexual abuse if not remembered will be repeated. The abusers or perpetrators will tend to be people who have been abused in their life but have not confronted or worked through that abuse. The abuse of another is a way of repeating that which is otherwise too unbearable to face. In this film, and it is what makes this film significant, the two main characters, like Oedipus searching for truth, will face their truth – will know the origins of their pain, their 'otherness', the very nexus of their being. One of my students studying this film in light of her own experience of being abused writes:

> people or little children whose
> 'fear' instinct had been exploited, where the child has been robbed of a
> will of his own, are left going after and ending up in a death roll in order
> to be devoured. The perpetrators need to complete the
> trajectory of this pain in them – to the point of total
> objectification – it's a repeat of the pattern – to find
> out at the edge of death – whether they would be let go of life,
> or be cannibalized.
> It appears that 'in-between' is just torture, the person
> who makes it to adulthood – is trying to answer this
> fundamental question about his or her own worth – their core value –
> terrified in truth that the perpetrator wants to devour them,
> cannibalize them, so instead they become their own cannibal –

> cannibal of their spirit – trying to find that truth – trying to answer the question: Am I worth the searching or do I keep bleeding out and dying slowly?
> Just like a flower or blade of grass outside me now – I say
> and see – there is no one to ask here whether the life
> experienced as a blade of grass or a flower – is worth
> anything. We are all worth our own self-possession, this 'I',
> that needs no measure – you can't measure worth, just as the
> flowers grow and the trees and the blades of grass – their
> meaning goes unquestioned, their worth is its essence, its
> own distinct self [...]

Where does the victim find this truth? There is choice beyond the status given by the abuser: there is choice involved.[2]

*Mysterious Skin* traces the journey of two abused children into adulthood to a position where real choice is possible. One of them remembers but embraces the guilt, as if personally deserved, and proceeds to live out a life of emotional non-engagement: as a gay hustler he allows and needs his body to be devoured. The other has no memory other than ominous flashes of disturbing enigmatic scenes; he searches for a lost five hours. Both characters were victims to their athletics coach in their childhood. The character who remembers holds the 'truth' for the other but, as a non-engaged psychotic person, choosing a life that allows his childhood abuse to be repeated endlessly, as well as showing signs of becoming a perpetrator of the crime, he has no interest in providing answers for the other.

The fact that he comes to be the means by which the second character is able to access his trauma is the turning point in the film. This occurs in a scene where he is asked to express physical tenderness rather than merely a sexual act. The scene is set in a beautifully ordered primarily white room. He has been paid by an AIDS sufferer to caress him rather than to provide an outlet for sexual desire. The use of the camera in this scene is particularly important; the employed 'gay hustler' is represented differently; his facial expressions take on an intensity of feeling rather than disengagement. The choice of the painting of 'The Girl with the Pearl Earring' that hangs over the bed is not an accidental one. The ways in which art is an artifact of consciousness is pervasive on many levels here, asking the viewer perhaps to contribute to the unraveling of the puzzle of what is remembered and what is forgotten, in art, as much as in life.

---

[2] Danielle H, letter to Ann McCulloch, May, 16, 2006.

The art of tragedy is primarily concerned with choices – the choice that the protagonist makes will be one that not only re-evaluates prevailing values but will also bring about his or her end. In *Mysterious Skin* the characters are brought to the point where at last there is some kind of choice because at least they are seeing reality for not only what it is but also how it came to be that way. The film is presented as much as an argument as it is a series of disturbing images. The thesis of the film is clear: abused children feel unworthy, will be outsiders, and some will repeat what was done to them compulsively. Whether or not confronting the original pain will bring about change for the victim remains unanswerable, though the preference for unlocking memories instead of being blocked unhealthily from them appears compelling.

The two protagonists find each other and together unlock some hidden doors into their embryonic selves within which their potential selves have been frozen and deprived of the right to live. The film ends with their insight into an unbearable truth. The repression has been lifted and the time has come for choices in life to be possible. The characters at the end of the film have broken the need or possibility of the Freudian compulsive repetition. They are now represented as remembering. It would be at this point that these characters would be in a position to undergo Nietzsche's psychological test. Their liberation must entail accepting their pain and not wishing their life other than it has been. It is, as Nietzsche expressed, a 'nauseous' situation, but it is one that will lead to the only kind of liberation that is possible.

On a personal level the acceptance of one's burden, the acknowledgement of one's pain and the ability to accept, not repress it, or turn it into resentiment, vengeance and bitterness, is the moment of release, it is the moment of triumph over one's pathology. In tragedies, whether in life or art, the story ends in death (actual or the end of that identity) and new moral orders or identities appear in new states of becoming. The cycle begins once more. Nietzsche, however, reminds us that such orders are constructed ones and have only provisional truths. Traditional tragedies representing the return of a moral order have long passed into history. Current representations of tragic themes are less inclined to create solutions but, like their forerunners, the moment that characterizes the experience of eternal recurrence will not be codified or turned to stone as a table of value; it will slip, slide and be forgotten, only again to appear; this is its power and the condition for the possibility of this power, is the forgetting that it has occurred before.

The ‘forgetting’ may be represented in art by what is referred to in *The Birth of Tragedy* as the Apollonian ‘veil of illusion’; the Dionysian insight has an intensity that must dissipate or be made bearable. This forgetting is however not the same as what is forgotten in the repressed person. The repressed person who cannot remember, but who is nevertheless formed by that which is buried, repeats in his or her behaviour compulsively the pain of what is not accepted or known to consciousness. The person experiencing the power of eternal recurrence is shown the entirety of their life and is led, however momentarily, to accept and not wish otherwise every aspect of that life, its misery and its joy, and proclaims in affirmation that it will return again and again. The rather grandiose title of the triumph over pathology that I have given this experience is that moment when the most familiar pain each of us has in relation to something in our lives comes again. However, in the moment of celebration of its return there is no longer a need to make it otherwise, to seek vengeance against anyone or to think that it can ever be other than this experience of suffering. One accepts the suffering as one’s own and momentarily is given such acceptance of it that one touches the experience of the possibilities promised to the ‘overman’. The greatness of tragedy is held in this knowledge that arrives fleetingly – it is the acceptance and celebration of a moment of power when one has no need to make it other. It is the moment when Oedipus gouges out his eyes, or as Walter Stein says of Lear, ‘that he has at least learned to live even in affliction and heartbreak and death’ (Stein 1969: 160–1). And it is in each of us, when some familiar pain returns to us of some rejection, loss or abuse, that we receive it not as an insurmountable burden but instead as our burden, precious in its claims, necessary in its impact. We become like Nietzsche when he stumbles on a moment of what he calls ‘sinful happiness’:

> To spend one’s life amid delicate and absurd things; a stranger to reality; half an artist, half a bird and metaphysician; with no care for reality, except now and then to acknowledge it in the manner of a good dancer with the tips of one’s toes; always tickled by some sunray of happiness; exuberant and encouraged even by misery – for misery *preserves* the happy man; fixing a little humorous tail to the holiest of things.
> (From *The Will to Power*, quoted in Lingis 1997: 61)

Or, as Nietzsche expresses it elsewhere when concerned with the return of reactive forces, there is the vision of a young shepherd, '[w]rithing, gagging, in spasms, his face distorted', with a heavy black snake hanging out of his mouth. When he becomes aware that the eternal return, that being of becoming, cannot be a nihilistic becoming, the shepherd is able to affirm the eternal return. One must bite off and spit out the snake's head. Then the shepherd is no longer man or shepherd: he 'was transformed radiant, laughing! Never on earth has a human being laughed as he laughed.'

Another becoming, another sensibility: The Overman.

## *References*

Allison, David B. (ed.). *The New Nietzsche*. London: The MIT Press, 1997.

Assoun, Paul-Laurent. *Freud and Nietzsche*. London: The Athlone Press, 2000.

Belliotto, Raymond Angelo. *Stalking Nietzsche*. London: Greenwood Press, 1998.

Deleuze, Gilles. *Difference and Repetition*. Transl. Paul Patton. New York: Columbia University Press, 1994.

Derrida, Jacques. 'The Question of Style.' *The New Nietzsche: Contemporary Styles of Interpretation*. Ed. David B. Allison. Cambridge, Mass: MIT Press, 1997. 107–120.

Eagleton, Terry. *Sweet Violence: The Idea of the Tragic*. Oxford: Blackwell Publishing, 2003.

Freud, Sigmund. 'Mourning and Melancholia.' *Standard Edition*. Vol. 14. Translated and edited by James Strachey. London: Hogarth Press, 1958. 243–58.

—. *The Psychopathology of Everyday Life. Standard Edition*. Vol. 6. Transl. and ed. James Strachey. London: Hogarth Press, 1958.

—. 'Remembering, Repeating, and Working Through.' *Standard Edition*. Vol. 12. Translated and edited by James Strachey. London: Hogarth Press, 1958. 147–56.

—. 'The 'Uncanny.' *The Norton Anthology of Theory and Criticism*. Vincent. B. Leitch (ed.). New York: W. W. Norton & Company, 2001.

Grosz, Elizabeth. *The Nick of Time: Politics, Evolution, and the Untimely*. Australia: Allen & Unwin, 2004.

Klossowski, Pierre. 'Nietzsche's Experience of the Eternal Return.' *The New Nietzsche: Contemporary Styles of Interpretation*. Allison David B. (ed.). Cambridge, Mass: MIT Press, 1997. 107–120.

Lingis, Alphonso. 'The Will to Power.' *The New Nietzsche*. Allison, David B (ed.). London: The MIT Press, 1997. 37–63.

Nietzsche, Friedrich. *Thus Spake Zarathustra: A Book for All and None*. Transl. Walter Kaufman. New York: Vintage, 1965.

—. *Beyond Good and Evil*. Transl. R. J. Hollingdale. London: Penguin Books, 1990.

—. *On the Genealogy of Morals* and *Ecco Homo*. Transl. Walter Kaufman. New York: Vintage, 1967.

—. *The Will to Power*. Transl. Walter Kaufmann and R. J. Hollingdale. New York: Vintage, 1968.

—. *The Gay Science*. Transl. Walter Kaufmann. New York: Vintage, 1974.

—. *The Birth of Tragedy and Other Writings*. Transl. Ronald Speirs. Cambridge: Cambridge University Press, 1999.

Ricoeur, Paul. *Memory, History, Forgetting*. Trans. by Kathleen Blamey & David Pellauer. Chicago: The University of Chicago, 2004.

Schama, Simon. *Hang-Ups: Essays on Painting (Mostly)*. London: BBC Books, 2004.

Stein, Walter. *Criticism as Dialogue*. Cambridge: Cambridge University Press, 1969.

## Notes on Contributors

*Dirk de Bruyn* teaches animation and digital culture at Deakin University.

*Khadija Z Carroll* works as a curator of Modern and Oceanic Art in Harvard Art Museums, and is a PhD student in Art History at Harvard University.

*Barry Empson* teaches German at the University of Otago.

*Louise Fairfax* completed a PhD at the University of Melbourne in which she applied a systems-theoretical approach to a selection of Goethe's works.

*Ron Goodrich* teaches in the School of Communication & Creative Arts at Deakin University.

*Ewen Jarvis* is a PhD student and sessional staff member at Deakin University.

*Uli Krahn* is a PhD student in the Department of English at the University of Sydney.

*Peter Leech* teaches Art History and Theory at the University of Otago.

*James McArdle* is Head of School of Visual Arts and Design at La Trobe University.

*Ann McCulloch* is Associate Head of School (Research), School of Communication & Creative Arts at Deakin University.

*Tim Mehigan* is Professor of Languages and Chair of the Department of Languages and Cultures at the University of Otago.

*Paul Monaghan* lectures in the School of Creative Arts (Theatre Studies) at the University of Melbourne.

*Kim Roberts* is a practising architect. She completed an Honours degree in Literature at Deakin University.

*Rose Woodcock* lectures in animation and digital culture at Deakin University.